MYSTERY ANIMALS OF THE BRITISH ISLES
Staffordshire

Nick Redfern & Glen Vaudrey

Typeset by Corinna Downes
Edited by Nadia Novali
Cover and Layout by PoppyPetal for CFZ Communications
Using Microsoft Word 2000, Microsoft Publisher 2000, Adobe Photoshop CS.

First published in Great Britain by CFZ Press

**CFZ Press
Myrtle Cottage
Woolsery
Bideford
North Devon
EX39 5QR**

© CFZ MMXIII

All rights reserved. Without limiting the rights under copyright reserved above, no part of this publication may be reproduced, stored in or introduced into a retrieval system, or transmitted, in any form of by any means (electronic, mechanical, photocopying, recording or otherwise), without the prior written permission of both the copyright owners and the publishers of this book.

ISBN: 978-1-909488-11-3

Dedication

To the many and varied good folk of Staffordshire who have been lucky enough (or, perhaps even, unlucky enough!) to have encountered a creature of the truly strange kind: this book is for all of you

Nick Redfern (2013)

Dedication

Dedicated to my beautiful wife Kerry

Glen Vaudrey (2013)

Books by Nick Redfern

A Covert Agenda
The FBI Files
Cosmic Crashes
Strange Secrets
Three Men Seeking Monsters
Body Snatchers in the Desert
On the Trail of the Saucer Spies
Celebrity Secrets
Memoirs of a Monster Hunter
Man-Monkey
There's something in the Woods
Science Fiction Secrets
Contactees
The Mystery Animals of the British Isles: Staffordshire

Books by Glen Vaudrey

Mystery Animals of the British Isles: The Northern Isles
Mystery Animals of the British Isles: The Western Isles
Sea Serpent Carcasses Scotland: from the Stronsa Monster to Loch Ness

Contents

Foreword	7
Introductions	9
Chapter I: *Beware the Black Dog*	13
Chapter II: *Water-Maidens*	29
Chapter III: *Out-of-Place Cats*	37
Chapter IV: *Creatures on the Loose*	41
Chapter V: *Staffordshire Goes Big-Cat Crazy!*	51
Chapter VI: *From Where Do The Cats Come?*	59
Chapter VII: *The Taigheirm Terror*	65
Chapter VIII: *Creatures of the Castle Ring*	75
Chapter IX: *Where the Bigfoot Lurks*	81
Chapter X: *Sasquatch Mania*	91
Chapter XI: *There's Something in the Water*	113
Chapter XII: *Monsters of the Full Moon*	125
Chapter XIII: *Wallaby Wonders*	139
Chapter XIV: *Boar to be Wild*	145
Chapter XV: *Alien Animals*	151
Chapter XVI: *Flying Monsters*	157
Chapter XVII: *Bad Bunnies*	167
Conclusions	171
Spotter's Guide	173
About the Author - Nick	189
Nick's References and Resources	191
Nick's Acknowledgements	196
About the Author - Glen	197
Glen's Acknowledgements and Bibliography	197
Index	199

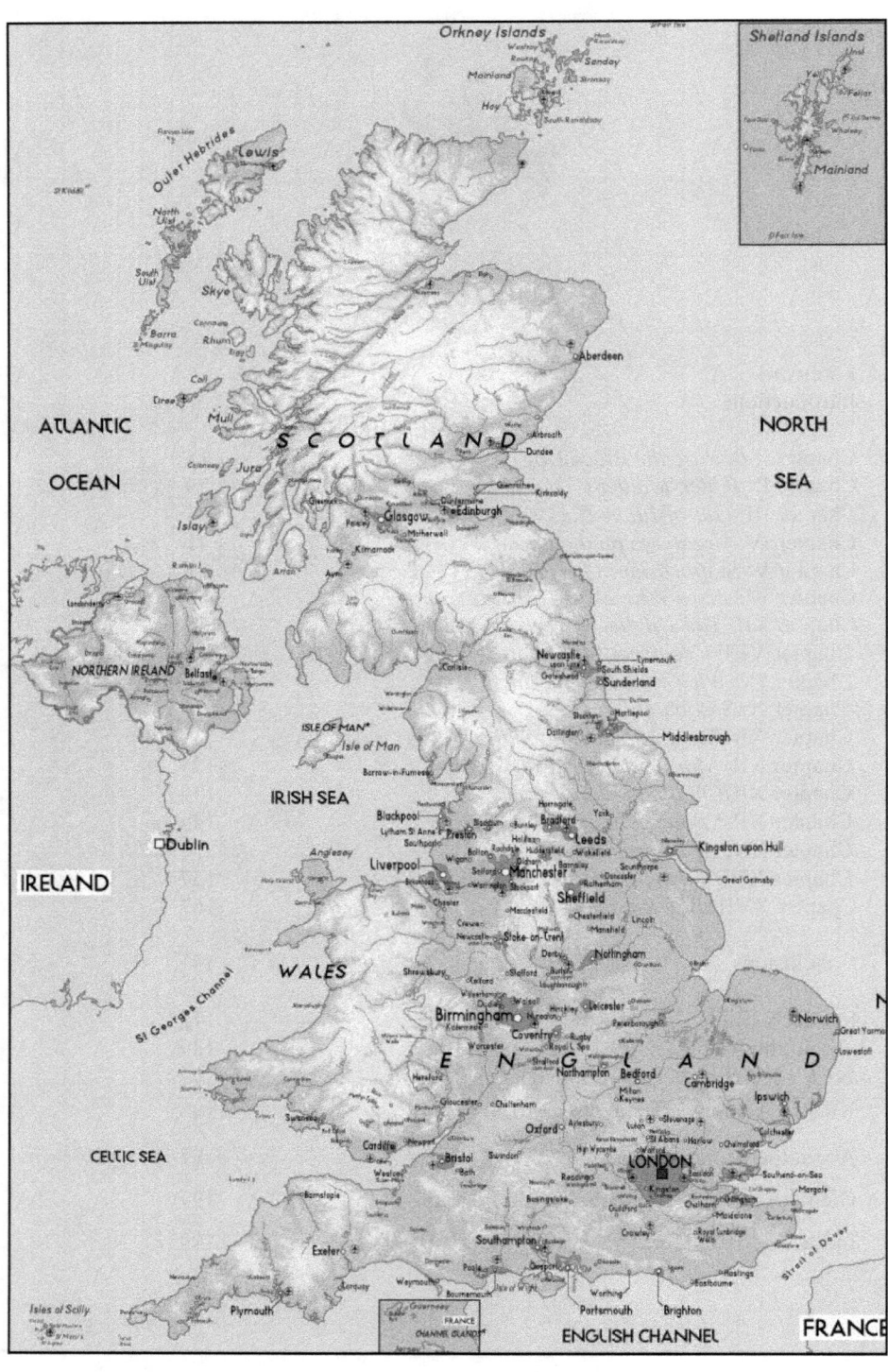

The Mystery Animals of the British Isles

More years ago than I care to remember, my first wife bought me a birthday present. It was a book about the mystery animals of Britain and Ireland, and I devoured it heartily. When I finished, I was horribly disappointed. It had covered the mystery cats of the country in some depth, as it had done with the black dog legends, and a smattering of more arcane 'things' (as the late, great Ivan T. Sanderson would doubtlessly have dubbed them) such as the Owlman of Mawnan, and the Big Grey Man of Ben Macdhui. But there was *so* much that I knew the author had simply left out.

Where were the mystery pine martins of the West Country? Where were the Sutherland polecats? Where was the mysterious butterfly known as Albin's Hampstead Eye? This was an Australian butterfly, the type of specimen which was caught in a cellar in Hampstead (hence the name) but no-one knows how or why. Where were the butterflies, moths, birds and even mammals known from the British Isles on the basis of a handful of specimens only? And where were the local oddities; the semi-folkloric beasts only known from a specific location?

Although at the time I had no pretensions to being a writer, I started to collect information from around the country; and, with the benefit of hindsight, it is probably with my disappointment with my 27th birthday present that the seeds of what would eventually grow into the Centre for Fortean Zoology, were planted.

Nearly twenty years later to the day, I was sat in my garden at the Centre for Fortean Zoology [CFZ] in North Devon, sharing a bottle of wine with my wife, and Corinna and my old friends Richard Freeman and Mike Hallowell. The subject of my disappointing 27th birthday present came up, and someone suggested that we do our best to redress the balance. CFZ Press, the publishing arm of the CFZ, has become the largest dedicated Fortean zoological publishers in the world,

The Mystery Animals of Staffordshire

and we are now in the position to put my vague daydreams of a couple of decades ago into action. We decided that rather than trying to publish one enormous tome covering the mystery animals of the whole of the British Isles (which, by the way, geographically, if not politically, includes the Republic of Ireland, but excludes the Channel Islands) we would be much happier presenting this vast array of data in a series of books, each covering a county or two. Then we realised the enormity of what we were proposing: The series would probably end up being something in the region of forty volumes in length!

However, never ones to back away from a challenge, we decided to go ahead with the project, and now – six months later – the first books in the series are being published.

We argued the toss for months over how we were going to format the series. For a long time we were intending to have a rigid format for all the books, something akin to the Observer's book of the British countryside. But then we decided 'No'. There are as many kinds of researcher as there are mystery animal, and it would – we felt - be more in keeping with the ethos of the CFZ, if we allowed each researcher to present his or her findings in their own inimitable style. The books, therefore, will reflect the character of the individual author.

Some will be poetic verging on mystical. Some will be matter of fact scientific. Some will be from the point of view of a naturalist, and some from the point of view of a folklorist. Some will be short, some will be long. Some will be full of scientific theorising, and some full of metaphysical speculation. But one thing is sure: whoever gets one of these volumes for their 27th birthday…

…They won't be disappointed!

Jonathan Downes
Director, Centre for Fortean Zoology
Woolfardisworthy
North Devon

Introduction

When the esteemed, respected and exceptionally-talented (he really *did* ask me to describe him that way, and I was handsomely rewarded in return) editor of this book, and the head-honcho of the Centre for Fortean Zoology to boot, Jonathan Downes asked me via a transatlantic telephone-call late one cold and icy winter's night in 2007 if I would be interested in writing the latest volume in his *Mystery Animals of the British Isles* series, I quickly and enthusiastically said: 'Yes!' And, for a couple of key and integral reasons:

First and foremost, when Jon initially told me of his grand, and admittedly somewhat daunting, plans to publish a comprehensive series of books that would detail the many strange beasts and out-of-place animals that have been reported throughout all of the various counties that collectively comprise the British Isles, I immediately thought – and I still *do* think – that it was a truly excellent idea.

After all, I carefully reasoned (a) no-one had previously ever embarked upon such a truly ambitious and wide-ranging project; and (b) each and every title would amply, and undoubtedly, serve to demonstrate the overwhelmingly rich diversity of beasts – some of straightforward flesh-and-blood origins, and others of a far stranger nature – that inhabit the darkened, and sometimes the not quite so darkened, corners of the green and fair land that is known as the British Isles. And, anyway, Jon is my closest mate within the strange realm of Forteana; and so I could hardly say 'no' to his generous offer, could I? Well, I could have done; but then that just wouldn't be cricket, would it?

As for the second reason why I'm so pleased to be able to present you with this book: well, it's because, as some of you may know, I spent many-a-year living deep within the heart of Staffordshire itself. And, as a result of that particular factor, I have come to firmly appreciate that beneath the county's image and veneer of friendly and inviting normality, there are distinctly strange and diabolical things afoot in these parts – and particularly so after the sun has set, when the moon is full, and while the chill wind howls loudly and ominously.

Exotic big cats are said to roam Staffordshire's thick woods; ghostly black dogs with red, glowing eyes faithfully haunt and patrol its ancient roads and well-worn pathways; bloodthirsty werewolves are rumoured to be on the loose throughout the county; fantastic and vicious water-beasts lurk deep within its streams, rivers, lakes and pools; out-of-place wallabies, wild-boar, porcupines, coypu and armadillos have all been reported; and even the world's most famous hairy man-beast of all, Bigfoot, has been known to put in an appearance from time to time!

In other words, there is a very deep, and infinitely varied and rich vein of data to be comprehensively and definitively mined – and so that is precisely what I have done right here. As a result, throughout the pages of the book that you now hold squarely in your hands, you will find startling, illuminating and occasionally terrifying stories of the mysterious animals and the beastly oddities that inhabit mysterious Staffordshire.

For me personally, having grown up in the county itself, but now living on the other side of the world (just outside of the city of Dallas, Texas), it was an absolute joy to be able to return to my old, green and wooded stomping grounds for a period of approximately five months spread across the last three years, and to secure a wealth of witness testimony, photographs and much more on the weird and wonderful wildlife that resides in the heart of mysterious Staffordshire.

There is one thing that I need to carefully stress from the outset: I make no apology whatsoever for the fact that many of the accounts that are revealed and related in the pages of this book originated from within the dense and mysterious woods of the Cannock Chase – as well as within its immediate, surrounding area, hamlets and villages. As seasoned students of the world of the unexplained will be acutely well aware, the Cannock Chase itself has been a veritable hotbed of absolute high-strangeness for many a year; and, therefore, it is only to be expected that reports from that particular locale should dominate these pages. I have, of course, been very careful to ensure that the rest of Staffordshire is justly and rightly covered, too!

And with all that now said, allow me to acquaint you with that particularly bizarre menagerie of monsters and mysterious animals that for whom, and once like me, Staffordshire can well and truly be called 'home'…

Nick Redfern, June 2013

Introduction

So how did I end up writing about Staffordshire?

Well to put it simply I slipped over the Cheshire border while no one was looking. For a number of years I have been busy researching information for the *Mystery Animals of Cheshire* book and whilst in the process of that research I happened to find some great tales from the south of the county and followed them only to find they were in Staffordshire these days because of boundary changes in the 1970s, rather than waste the information I managed to find a home for it in this fine volume.

My input can be found popping up at the end of chapters and I hope that you find my writing every bit as interesting as Nick's.

Glen Vaudrey, June 2013

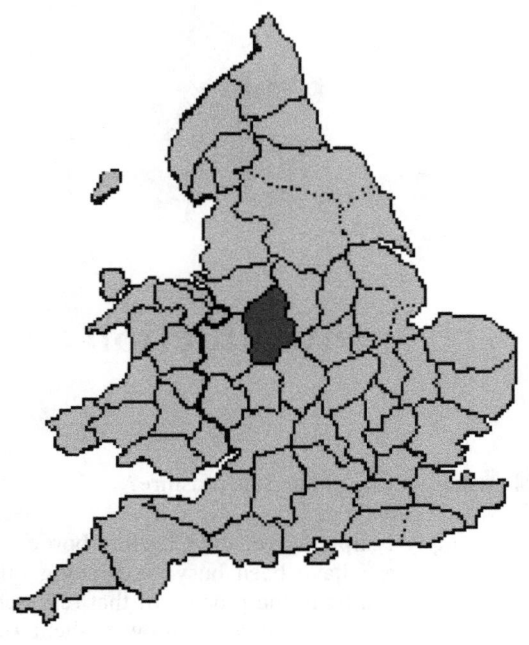

Chapter I
Beware the Black Dog

'...a "wolf-like" creature was spotted by dozens of motorists on the M6 hard-shoulder...'

In his absolutely definitive book *Explore Phantom Black Dogs*, the author and researcher Bob Trubshaw wrote the following: 'The folklore of phantom black dogs is known throughout the British Isles. From the Black Shuck of East Anglia to the Mauthe Dhoog of the Isle of Man there are tales of huge spectral hounds "darker than the night sky" with eyes "glowing red as burning coals". The phantom black dog of British and Irish folklore, which often forewarns of death, is part of a world-wide belief that dogs are sensitive to spirits and the approach of death, and keep watch over the dead and dying. North European and Scandinavian myths dating back to the Iron Age depict dogs as corpse eaters and the guardians of the roads to hell. Medieval folklore includes a variety of "Devil dogs" and spectral hounds.'

One of the most infamous of all black-dog encounters in the British Isles occurred at St. Mary's Church, Bungay, Suffolk, England, on Sunday, August 4, 1577, when an immense and veritable spectral hound from hell materialised within the church during a powerful thunderstorm and mercilessly tore into the terrified congregation with its huge fangs and razor-sharp claws. In fact, so powerful was the storm that it reportedly killed two men in the belfry as the church tower received an immense lightning bolt that tore through it and shook the building to its ancient foundations.

According to an old, local verse:

> 'All down the church in midst of fire, the hellish monster flew.
> And, passing onward to the quire, he many people slew.'

Then, just as suddenly as it had appeared, the beast bounded out of St. Mary's and was reported shortly thereafter at Blythburgh Church, about twelve miles away, where it allegedly killed and mauled even more people with its immense and bone-crushing jaws – and where, it is said, the scorch marks of the beast's claws can still be seen to this day, infamously imprinted upon the ancient door of the church.

Even more intriguing is the fact that Bungay's legend of a satanic black hound parallels that of yet another local legend: that of Black Shuck, a giant, spectral dog that haunts the Norfolk and

Suffolk coasts. Indeed, such is the popularity of the Bungay legend, that it has resulted in an image of the beast being incorporated into the town's coat of arms - and the Black Dogs is the name of Bungay Town Football Club.

The stark, disturbing and memorable image that the infamous devil dog, or the phantom hound, as described above undoubtedly conjures up is that of a definitively sinister beast that stealthily prowls the towns and villages of ye olde England by nothing more than silvery moonlight or to the accompanying background of a violent, crashing thunderstorm. It is, however, a little known fact outside of dedicated students of the phenomenon that sightings of such creatures continue to surface right up until modern-day times.

One particular area which seems to attract far more than its fair share of encounters and reports of the paranormal-pooch variety, is that aforementioned sprawling mass of dense forest in Staffordshire known as the Cannock Chase. Among the folk of the many small villages that sit on the fringes of the Cannock Chase – or that, in some cases, can be found deep within the very heart of its shadowy, wooded depths – terrifying tales of the diabolical hounds of hell are disturbingly, and surprisingly, common. In other words, the traditions and the beliefs of times now-long-gone have actually not faded away at all. In fact, the precise opposite is the case: they are still firmly and widely among us – and are, somewhat secretly, even embraced, too.

A high plateau that is bordered by the Trent Valley to the north and the county of the West Midlands to the south, the sprawling Cannock Chase has been a highly cherished and integral feature of the Staffordshire landscape for countless generations. Following its initial onslaught on the British Isles in A.D. 43, the armed-forces of the Roman Empire headed to the south - to what is now the town of Cannock - and along a route that later became known as Watling Street: a major, and now-historic, Roman road. It can be amply demonstrated that the surrounding countryside was very heavily-forested even way back then: for example, the Romans' colorful name for the nearby village of Wall was *Letocetum*, or, to give it its modern day translation: the Grey Woods.

In 1189, King Richard I granted the Cannock Chase as a hunting reserve to the then-Bishop of Lichfield; and, as far back as 600 years ago, the area of the Chase known today as Brockton Coppice was already overflowing with hundreds of sessile oak-trees. Today, some of those original, mighty oaks that were seeded all those centuries ago are still said to stand, both proud and tall alike. Little wonder, therefore, that in 1958, the Cannock Chase was designated by the British Government of the day as an Area of Outstanding Natural Beauty (AONB).

And, with that all now said, let me firmly acquaint you with Staffordshire's hounds of a truly monstrous kind.

Late one evening in the early weeks of 1972, a man named Nigel Lea was driving blissfully across the Cannock Chase when his attention was suddenly drawn to a strange ball of glowing, blue light that seemingly came out of nowhere and slammed violently into the ground some short distance ahead of his vehicle and amid nothing less than a veritable torrent of bright, fiery sparks. Needless to say, Lea quickly slowed his car down to what was a literal snail's pace. And, as he cautiously approached the approximate area where the light had seemingly fallen, was both shocked and horrified to the absolute core to see looming before him, 'the biggest bloody dog I have ever seen in my life'.

Very muscular and utterly black in colour, with a pair of large, pointed ears and huge, thick paws, the creature seemed to positively ooze both extreme menace and overpowering negativity, and had a crazed, staring look in its yellow-tinged eyes. For twenty or thirty seconds or so, both man and beast alike squared off against each other in classic stalemate fashion, after which time the 'animal' both slowly and carefully headed for the darkness and the camouflage of the tall, surrounding trees, not even once taking its penetrating eyes off of the petrified driver as it did so.

Somewhat ominously it might be said, and only around two or three weeks later at the most, says Lea, a very close friend of his who he had known since his earliest schooldays was killed in an industrial accident, under very horrific circumstances, in the West Midlands town of West Bromwich; something which, today, Lea firmly believes – after having deeply studied, almost to the point of total obsession, the history of British Black Dog lore and the creature's associations with tragedy and death – was directly connected with his strange and unsettling encounter on that tree-shrouded road back in 1972.

It is also interesting to note that this was not the only British-based phantom black dog encounter of 1972: on April 19 of that same year, at Gorleston, Norfolk for example, a coastguard named Graham Grant was witness to a huge black dog that was seen charging along the nearby beaches – until it vanished into thin air, that is, never ever to return.

Then, at around 3.20-3.30 a.m. on a particular day in the latter part of September 1972, a nurse returning home from a late-shift at a Sheffield hospital, encountered a glowing-eyed, ghostly dog briefly padding around – in what was perceived by the witness as a 'disturbed or confused' fashion – on her doorstep, before leaping off into the ethereal, autumn darkness.

Perhaps not merely a coincidence is the revelation that only hours earlier, the nurse had been directly involved in a tragically-unsuccessful attempt to save the life of a young car-accident victim who had been fatally injured in a head-on crash on a main road situated on the fringes of the city of Sheffield. In a somewhat highly synchronistic fashion, today the now-retired nurse and her husband make their home in the Staffordshire town of Hednesford.

In the early-to-mid 1980s, truly surreal and sinister reports began to surface of a creature that became known at a local level as the Ghost Dog of Brereton – a reference to the specific area of Staffordshire from where most of the sightings originated. Brereton once had its very own identity; but today it is considered to be a part of the town of Rugeley – or Rudgeley, as it was originally known, according to the *Domesday Book*, and which translates as 'the hill over the field'.

With specific respect to the Brereton encounters, the phantom dog at issue was described as being both large and frightening, and on at least two occasions it reportedly vanished into thin air after having been seen by terrified members of the public on lonely stretches of ancient road late at night. In direct response to an article that appeared in the *Cannock Advertiser* newspaper during the winter of 1984/5 on the sightings of Brereton's infamous ghost dog, a member of the public from a local village, one Sylvia Everett, wrote to the newspaper thus:

> 'On reading the article my husband and I were astonished. We recalled an incident which happened in July some four or five years ago driving home from a celebration meal at the *Cedar Tree* restaurant at about 11.30 p.m. We had driven up Coal Pit Lane and were just on the bends before the approach to the *Holly Bush* when, from the high hedge of trees on the right hand side of the road, the headlights picked out a misty shape which moved across the road and into the trees opposite.'

Mrs. Everett continued with her account:

> 'We both saw it. It had no definite shape seeming to be a ribbon of mist about 18in. to 2ft. in depth and perhaps nine or 10ft. long with a definite beginning and end. It was a clear, warm night with no mist anywhere else. We were both rather stunned and my husband's first words were: "My goodness! Did you see that?" I remember remarking I thought it was a ghost. Until now we had no idea of the history of the area or any possible explanation for a haunting. Of course, this occurrence may be nothing to do with the 'ghost dog' or may even have a natural explanation. However, we formed the immediate impression that what we saw was something paranormal.'

Another person who may very well have seen the phantom hound of Brereton was Sally Armstrong. It was shortly after the breaking of dawn one day in late March 1987, and Armstrong, a now-retired employee of a Shropshire-based auctioneering company, was on her way to meet with a client, then living in Brereton, who was employed in the antiques trade. For a while at least, all was completely and utterly normal. But, things were only destined to change - and for the absolute worst, too, it can be convincingly argued.

Shortly before she arrived at the old cottage of the man in question, Armstrong was witness to a monstrous black-hued dog with wild, staring eyes that was sitting at the edge of the main road that runs through the locale of Brereton, and which was staring intently at her as she passed by it. Somewhat unsettling: as Armstrong drove by the huge beast, she slowed down, quickly looked in her rear-view mirror, and could see that its head had now turned in her direction. It was, apparently, still focusing upon her each and every move.

Armstrong concedes that there was nothing to definitively suggest an air of the supernatural or the paranormal about the fiend-dog she saw nearly a quarter of a century ago; however, that its huge presence seemed to both surprise and unsettle her for reasons that she cannot to this day readily explain or rationalise properly, leads Armstrong to conclude that: 'there was just something about it that makes me remember it this much later'.

Possibly of deep relevance to the tale of the ghost dog of Brereton was the story of a man named Ivan Vinnel. In 1934, as a twelve-year-old, he had a very strange encounter indeed in

The Mystery Animals of Staffordshire

his hometown of nearby Burntwood. The sun was beginning to set and the young Ivan and a friend were getting ready to head home after an afternoon of playing hide-and-seek. Suddenly, however, the pair was stopped dead in its tracks by the shocking sight of a ghostly 'tall, dark man', who was 'accompanied by a black dog' that had seemingly materialised out of a 'dense hedge' situated approximately ten-yards from the boys' position. Both man and animal passed by in complete and utter silence before disappearing – in typical and classic ghostly fashion, no less.

Ivan later happened to mention the details of the unsettling incident to his uncle, who then quietly and guardedly proceeded to tell him that he, too, had actually seen the ghostly dog on several occasions when he was a young child. And, as is typically the case with ghostly hounds all across the British Isles, the beast was always reportedly seen in the same location: namely, faithfully pacing along the old road that stretches from the village of Woodhouses to an area of Burntwood situated near the town's hospital. And: weird reports from Staffordshire's out-of-place, dog-like beasts continue to surface to this very day.

It was in the latter part of June 2006 that all hell metaphorically broke loose, when reports flew wildly around the town of Cannock to the effect that nothing less than a fully-grown wolf was roaming and rampaging around the area. Early on the morning of June 28, motorists on Junction 10A of the M6 Motorway near Cannock jammed Highways Agency telephone-lines with reports of a 'wolf-like creature' that was seen 'racing between lanes at rush hour'. Gob-smacked motorists stared with complete disbelief as the immense beast, described as being 'grayish-black', raced between lanes, skilfully dodging cars, before leaping for cover in the nearby trees.

Highways Agency staff took the reports very seriously at the time, but publicly concluded that the animal was most 'probably a husky dog'. However, a spokesperson for *Saga Radio* – which was the first media outlet to arrive on the scene – said in reply to the statement of the Highways Agency that: 'Everyone who saw it is convinced it was something more than a domestic dog. I know it sounds crazy but these people think they've seen a wolf.'

The local newspaper, the *Chase Post*, which has always been *very* quick off-the-mark to report on incidents of mystery animals seen in the vicinity of Staffordshire's old woods, stated on July 6 in an article titled *Great Beast Debate on Net* that: 'Internet message boards are being flooded with debates on our front-page revelation last week that a 'wolf-like' creature was spotted by dozens of motorists on the M6 hard-shoulder.'

The *Chase Post* further noted, with justified pride and perhaps even a little welcome surprise, that: 'Our own website has been thrown into overdrive by the story, which received around 2,600 hits from fans of the unexplained across the globe in the last week alone.'

While the highly mystifying affair was certainly never ultimately resolved to the satisfaction of everyone involved – or, it might reasonably be said, to the satisfaction of *anyone* involved, for that matter - the final words went to the Highways Agency, a spokesperson for whom stated that: 'We have received a number of reports that the animal was captured. But we don't know where, who by, or what it was.' Of course, this truly open-ended and somewhat vague statement did nothing to resolve the matter, at all.

Perhaps the event had indeed been due to the mistaken sighting of an escaped Husky; however, that does not in any way come close to explaining the eerie encounter of Jim Broadhurst and his wife that occurred while the pair was out for a morning stroll on the Cannock Chase, only a matter of days before the memorably-monstrous events of June 28, 2006 took place.

Broadhurst states that he and his wife had seen at a distance of about one hundred and fifty feet, what looked very much like a large wolf or 'a giant dog' striding purposefully through an opening in the woods. Broadhurst added that deep fear firmly gripped the pair when the creature suddenly stopped and looked intently and menacingly in their direction. That fear was amplified even further, however, when the beast reportedly, and bizarrely, reared up onto two powerful hind legs and backed away into the thick trees, never to be seen again. The husband and wife, unsurprisingly, fled those dangerous and eerie woods – and have not returned since; ever-fearful of what they believe to be some form of 'monster' lurking deep within the mysterious depths of the Cannock Chase.

Interestingly, and certainly unfortunately, in the weeks that followed the encounter, the Broadhurst family was cursed with a seemingly never-ending run of bad luck and disaster that did not abate until well into September of that same year.

Moving away from the Cannock Chase, there is the story of the Bradley family of Leeds who had the very deep misfortune to encounter one of the now-familiar hounds of hell in early 2009: at no less a site than the Staffordshire city of Lichfield's famous and historic cathedral; which has the distinction of being the only English cathedral to be adorned with three spires.

According to the Bradleys, while walking around the outside of the cathedral one pleasant Sunday morning, they were startled by the sight of a large black dog 'barrelling' along at high speed, and adjacent to the side of the cathedral. The jaw-dropping fact that the dog was practically the size of a donkey ensured their attention was caught and held. But that attention was rapidly replaced by overwhelming fear, when the dog allegedly 'charged the wall' of the cathedral and summarily vanished right into the brickwork as it did so! Perhaps understandably, the Bradleys chose not to report their mysterious encounter to cathedral officials, to the police, or to local media outlets.

Interestingly, a very similar beast – if not, perhaps, the very *same* one - is believed to have been seen only a short distance from the cathedral by a Scottish family with the surname of either Dobson or Robinson way back in the early-to-mid-1950s. The details of this encounter are admittedly vague, scant and hazy in the extreme, however, and were passed on to me merely as an aside and nothing else really, in the autumn of 1997 by a now-retired journalist who was working in the area at the time.

Two further stories of ghostly hounds seen in the county of Staffordshire come from paranormal investigator Tim Prevett, the first of which is focused around Swythamley Hall; a late-18th Century country house that can be found near Leek, Staffordshire, and which, today, is classified as a Grade II listed building that has been converted into four separate residences. The manor of Swythamley was held by the Crown following the dissolution of Dieuclacres Abbey and, thereafter, had several owners. It was acquired by the Trafford family in 1654, who replaced the original manor-house with a new construction around 1690. The family remained in residence until Edward Trafford Nicholls – who was the High Sheriff of Staffordshire in 1818 - sold the estate to Sir Philip Brocklehurst in 1832.

And, of the particular ghost-hounds of Swythamley, Prevett says: 'One of the Traffords of Swythamley, while out hunting with dogs, is said to have leaped over a chasm. Having successfully cleared the precipice with his horse, the hunting hounds were not so lucky. They fell and perished in Lud's Church, where their spectral howls are still heard on occasion today. Their owner is also said to return.'

Prevett adds: 'A little way to the south west is Gun Hill, past which Bonnie Prince Charlie and his army passed in December 1745. The hill was also the site of gallows. A black dog haunts this spot, and elsewhere en route to Derby via Leek and Ashbourne, where members of the Highlanders' rebellion perished. It is said the black dogs either mark or guard the Jacobites' final resting places.'

Very notably, and somewhat curiously, Swythamley Hall plays a key and integral role in the strange story of yet *another* Staffordshire animal-based puzzle: namely that which surrounds the county's very own, and once-quite-large, population of out-of-place wallabies – an odd, yet enlightening, story that will be detailed in a later chapter.

And then we have the account of Marjorie Sanders. Although Sanders' account can be considered a new one – at least, in the sense that it only reached my eyes and ears in August 2009, during which time I was on a week-long return trip to England – it actually occurred back in the closing stages of the Second World War, when the witness was a girl of ten or eleven. At the time, Sanders was living in a small village not too far from Tamworth Castle – which overlooks the River Tame, and which has stood there since it was built by the Normans in the 11th Century; although an earlier, Anglo-Saxon castle is known to have existed on the same site, and which was constructed by the forces of Ethelfreda, the Mercian queen and the eldest daughter of King Alfred the Great of Wessex.

According to Sanders, 'probably in about early 1945', her grandfather had 'seen a hell-hound parading around the outside of the castle that scared him half to death when it vanished in front of him'. For reasons that Sanders cannot now remember or comprehend, her grandfather always thereafter memorably referred to the animal in question as 'the furnace dog'. Whether or not this is an indication that the spectral dog had the seemingly-ubiquitous fiery red eyes that so many witnesses have reported remains unfortunately unknown; but, it would not at all surprise me if that was one day shown to be the case.

Then, we have the brief but highly thought-provoking account of Gerald Clarke, a Glasgow baker, whose father claimed to have briefly seen a phantom hound with bright, electric-blue-coloured eyes on the grounds of Royal Air Force Stafford in the late 1950s, and while on patrol late one winter's evening. As was the case with so many other witnesses to such disturbing entities, the elder Clarke quietly confided in his son that the creature 'just vanished: first it was there and then it wasn't'.

What is particularly interesting about this case is that – as will become very apparent later – this was reportedly not the only occasion upon which an unusual, out-of-place creature was seen roaming around the secure grounds of RAF Stafford.

Finally, there is the following, very significant story, which surfaced quite literally as I was about to submit the original Word document of this book to the publisher, Jon Downes. In an article titled *Fresh Sighting of UFOs and Werewolves on Cannock Chase* that appeared in the *Birmingham Post* newspaper on January 15, 2010, it was reported that: 'Resident, Jane McNally, recently had a run-in with a mysterious canine creature while out walking with her partner on Cannock Chase.'

The *Birmingham Post* quoted McNally as saying: 'I was walking with my partner and his dog. We put the dog back on the lead as we thought in the distance there was an enormous dog. As we approached the animal we realised this wasn't a dog and it just stared at us for a while – I said it looked like a fox, but the size of a lioness – it then turned into the wooded area, and we proceeded to walk on. As it turned its long, bushy black tipped tail, we realised it was definitely not a dog. I have just logged onto the net and went on to images of wolves, and can honestly say whatever we saw yesterday was the closest thing to a wolf.'

And, now, with all of the above-cases both highlighted and digested, a final word or several on the overall black dog mystery is surely required: whether they are (a) the precursors to inevitable doom, tragedy and death, as the story of Nigel Lea *does* admittedly tend to suggest; (b) ancient and paranormal entities that are somehow connected with the realm of the dead and the afterlife; or (c) the spirits of long-deceased animals that have returned to forever watch over and guard the living – or even, perhaps, a strange and presently-unfathomable combination of all three theories - it seems that the phantom black dogs of Staffordshire are most assuredly here to stay...

Glen Vaudrey:

Well having read about all the dogs that Nick has managed to find it's now my turn to introduce you to some of the hounds that are to be found not only in and around Staffordshire but also occasionally in the air above. Yes in the air, I bet that has you wondering. Parachuting dogs? No, it's the Gabriel hounds.

It was in 1668 in Hammerwich that a man by the name of Francis Aldridge reported that he had heard a whistling noise up in the air 'the tune more melodious to him than any he had ever listened to in his life.' When, over two hundred years later, Charles Poole was writing about this incident he likened the noise Aldridge had heard to that of the Gabriel hounds. The sound of these aerial hounds was often taken as a portent to a disaster; considering the shocking safety standards in industry at the time it was hardly surprisingly that they could be tied in to any number of disasters.

Hammerwich wasn't the only place in Staffordshire where the Gabriel hounds were heard, they were also heard in Wednesbury. It was said that miners on the way to the pit often heard the sound of the hounds high above; as this seems to have happened on a regular basis perhaps it wasn't seen as the same portent of doom that it was elsewhere.

As for what the Gabriel hounds could be there are many theories, these range from the sound being that of angels transporting blessed souls off to heaven, or alternatively the Gabriel hounds have been ascribed to be the souls of infants who had died before being baptized and so were condemned to wander. The Gabriel hounds were not unique in being aerial hounds, most parts of the country have some form of the wild hunt, from the headless hounds that chase Jan Tregeagle across Cornwall to the fairy hounds of the *Sluagh* that race over the Outer Hebrides on their way to the land of the ever young, Tír-nan-Óg.

Not all hounds are aerial, there are some ghostly dogs to be found in Staffordshire that appear to plod around on the ground seemingly in order to just leave a trail of confusion and questions in their wake.

It was in 1913 that news of ghostly happenings from Hermitage Farm, Ipstones started to be reported in the papers worldwide, appearing as far apart as the United States and Australia. It is hard to imagine now that the haunted goings on in a farmhouse in Staffordshire would get such worldwide press coverage; then again with the internet I am sure it would get better coverage these days if put on YouTube. So what was happening a century ago that was so interesting? Well it seems that the tenant farmer, Bennett Fallows, started to report all kinds of hauntings at the 300-year-old Hermitage Farm. Fallows told how, during the thirteen years that he had been the tenant at the farm, his family had often seen a phantom hound. He recounted how one visitor had tried to kick the dog but as he went to kick it his foot simply went through the animal. I bet that caused a bit of a shock, but then why was he trying to kick it in the first place?

As well as the ghostly hound there was other strangeness to be found in the farm building. Fallows told tales of hearing the sound of people walking about, unearthly screaming, furniture moving about and the fact the doors no longer shut properly, well I wonder if that last one was a result of subsidence?

In 1974 Jean Haigh writing in *Country Life* told that on a stretch of road in the moorlands near Leek her grandfather had encountered a phantom dog one night. He wasn't the only family member to have an encounter, Jean's aunt also had a sighting of her own, she recalled that she once felt something sniffing her hand one evening whilst walking home, she turned around to find there was nothing behind her neither a phantom dog nor a dwarf hand-sniffer.

Just in case you're worried that every phantom dog comes with a body count that would keep the grim reaper busy for a month of Sundays let me reassure you that not every ghostly mutt behaves in this way, indeed some don't seem to serve any purpose that can easily be understood.

While there doesn't seem to be any good reason for the Leek dog mentioned above to be wandering about there is another report of a spectral hound in Staffordshire and this one is said to be accompanied by a ghost. The location of this sighting is at one of the region's best known tourist attractions, Alton Towers.

It was in December 1969 that the author Peter Travis went to Alton Towers to interview Mr Noakes, the foreman of the estate. Mr Noakes explained that he was not a man scared of the dark which was just as well really as his job involved walking around the grounds at various times of the day and night. While Mr Noakes didn't find anything amiss with these locations after dark Travis himself did comment on the fact that there were some areas that did have an eerie atmosphere.

The events that Mr Noakes told started quite mundanely. On the day, or rather night, in question he had just seen his fiancée onto the train at the southern side of the Towers (this station closed in 1964 thanks to Dr Beeching so that gives you an idea of when the sighting occurred). Having seen her safely onto the train Noakes set off back to the lodge along the 'step walk', a pathway that winds upwards through what was a heavily wooded part of the grounds, it has eighteen flights of steps, each flight consisting of around twelve steps and between each flight there was a space of around fifty yards. It was around 9.45pm when he made his way along this route and as he had walked it many times he wasn't expecting anything untoward but on this occasion something very different happened. Mr Noakes was approaching the last flight of steps when he saw someone standing at the top, around thirty yards from where he was. As he got closer to the figure he was not particularly disturbed by him as it was quite likely that the man was simply someone who had been visiting the house. As he got nearer to the figure it started to head down the steps towards him, Noakes noticed that the man was wearing a tall dark hat and a long flowing black cape, with a white scarf hanging loosely from his shoulders and in one hand he carried a silver topped black ebony cane. As the mysterious man drew level Mr Noakes said 'good evening' but as soon as he said the words the figure vanished and in spite of searching he could find no trace of this gentleman. When he arrived home he told his father of the sighting but his father showed skepticism.

The next day during a work break he was talking to some of his fellow workmates about the event of the previous evening, on the whole they didn't think anything of it suggesting that it could just be his imagination. However one old man who had worked on the estate for a number of years took it more seriously and asked if Noakes had seen a dog with the man, he replied that he had not. The older man then went on to tell him that he had seen the ghostly figure on many occasions and that on each time it had been accompanied by a black dog. The old man recalled that he had heard that there had been an accident years before when a guest from a party at the nearby Farley Hall had been heading back to the Tower but he never made it and the next morning his body was found at the top of the last set of steps. The cause of his death was unknown but foul play was not suspected. Since that date it appears that he has been haunting the site along with his dog. After Mr Noakes' initial sighting he would have only one further sighting of it and his once skeptical father would also have a sighting himself of both the man and the ghostly black dog.

From that last tale we see an association between death and the black dog, but the next mystery black dog we are going to look at took the theme to a whole new level; if the tales are to be believed its victims were legion.

We find this mysterious phantom hound in the region of Kidsgrove, a small town in north Staffordshire close to the border with Cheshire. Kidsgrove is also close to the Trent and Mersey

Canal and the rather impressive Harecastle Tunnel. At one time the appearance of a black dog was thought to herald trouble in the mines. Today the mines are gone and as tales of the black dog faded they would be replaced by stories of a ghostly boggart said to haunt Harecastle Tunnel and its environs. Before we look at reports of the phantom dog it would be as well to consider the role of the boggart in this story and to do that we need to know what a boggart is. So with that in mind let me introduce you to the boggart.

What a boggart is depends on what part of the country you are in; generally the definition is that of a folkloric hobgoblin sometimes reputed to be helpful while on other occasions it is described as being malicious. It really depended on how the boggart had been treated. While outdoors a boggart might have been happy to help a farmer with some back-breaking farming task, when indoors its presence was felt by the household by the number of small accidents or strange sounds that could be unaccountably heard. It is in the latter guise that the boggart is likened to a poltergeist. At other times when the boggart has been spotted it has been described as looking like a white cow, a white horse or a black dog. It might be fair to say that the boggart is whatever you want it to be.

A headless woman is said to haunt Harecastle Tunnel and its environs and her ghost has come to be known as the 'Kidsgrove boggart'. However as we shall see below the name probably belongs to something else entirely.

To digress for a moment Kidsgrove isn't haunted only by the boggart and the black dog, there are also stories of a ghostly white rabbit that could be found hopping around Clough Hall. You can read all about that spectral bunny elsewhere in this book.

As mentioned earlier, sightings of the dog were associated with the mines in the area, as too was the boggart and it wasn't in the form of a headless woman, this was before the name boggart became attached to the Harecastle Tunnel.

The first mention of the Kidsgrove boggart appears in the late eighteenth century as the 'Kitcrew Buggat', it may well have been reported before that date but it is with the growth of coalmines in the area that the buggat or boggart starts to appear more often.

One of the earliest accounts that links the dog and the boggart comes from the pen of Hugh Bourne who would go on to found Primitive Methodism alongside William Clowes, but back in 1816 he was just writing in his journal and in his entry for Tuesday 10th September of that year we find:

'I was employed in writing the commentary of the Bible adapted to the capacities of children. At night I was at Talk o' th' Hill and spoke from 'Moses' Bush'. I came to Kidcrew said properly to be Kidsgrove. They have informed me that several had been lately hurt in the coal works; limbs 'has' frequently been broken, though no lives lost. They also informed me that the Kidcrew buggat has been very much about. Something generally invisible that makes strange groans and dreadful noise, and stops near a certain house just before an accident happens in the family of that house. The buggat has lately been seen like a man barefoot and part undrest running swiftly after something like a white dog. When this invisible being has been heard about, the colliers always expect calamities to happen, and they say they uniformly happen.'

From this early report then we find that miners had already made the connection between a visit from the boggart and the accidents both underground and above. We also find out that trying to describe a boggart was harder than it should have been. When invisible its passing is known from the strange sounds that followed in its wake but if you think an invisible entity is a mystery then what do we make of the description of it being a half dressed man running after a white dog? I am not sure what can be made of that, is it a two-part creature or multiple boggarts? Then again it could just been a sighting of a bloke straight out of bed chasing a dog that might have stolen his breakfast, or then again a member of the odd squad on the prowl with unhealthy thoughts on his mind.

After this report of the boggart as a man-and-dog combination other events in the area rapidly overtook the true boggart and boggart Mark II was born, the headless woman. It wasn't because of changing tastes but rather it was as a result of a shocking murder that had taken place.

It was in 1839 that Christina Collins was travelling as a passenger on a narrowboat heading down the Trent and Mersey Canal. It wasn't the best journey as the crew were not the best, actually that's a very big understatement as by the time they reached the Rugeley aqueduct Mrs Collins was dead, murdered by the crew. Her body was found floating face down in the canal at Brindley Bank. Despite being found nowhere near Kidsgrove the events went on to influence the account of the Kidsgrove boggart as recorded by L.T. Meade.

It was in Meade's novel *Water Gipsies: Or, the adventures of Tag, Rag or Bobtail* published in 1879 that the late Mrs Collins was to be merged with the Kitcrew boggart. During the story an old bargee tells the tale of the tunnel being haunted by a woman who was murdered within it and who now haunts the area either as a white horse or as a headless woman. It is from this fictional account that the boggart seems to be known today, no longer does it appear as a half-naked man and a dog and after all why would it, a headless woman haunting the area sounds a much better attention grabber. The real murder victim kept her head unlike the ghost who certainly didn't, but that isn't the most surprising thing about this ghost, rather it's the fact that she was actually spotted by someone. In a collection of local folklore being collected in the 1950s there is an old account of Miss Napier who reported that she had had an encounter with the ghost of a headless woman dressed in old fashioned clothing. Does that mean she was influenced by Meade's writing or was Meade influenced by something already there, perhaps the boggart really was the inspiration.

All this talk of the boggart as a headless woman almost hid the boggart as a phantom dog but as we will now see the dog was rather busy foretelling disaster. Unlike the earliest report of it the dog is no longer white it has gone over to the dark side.

It is largely thanks to the work of Norman Roache, a local historian who in the 1950s started to record local folklore for stories and legends of the Kidsgrove area, that we have so many tales of the black dog. He recorded that many folk of the area had started to confuse the boggart and the black dog. The phantom hound was said to haunt the fields of Lower Ash which stretched across Newcastle Road, this area today is now a housing estate. Roache noted that many people 'have heard this phantom, but no one has ever seen it.' Is it me or does that make the idea of the phantom dog being black a little hard to believe? It could be that Roache understood that a black dog was better than an invisible dog, for a start it's easier to see.

Sightings of the black dog were persistent between 1860 and 1900. Roache recalled there were a number of mining disasters in the region during those years, so how do the sightings match up to mining accidents?

The Mystery Animals of Staffordshire

Without giving much in the way of details he mentions a sighting of the dog before the 1866 Talk o' th' Hill disaster when a gas explosion killed 91. While that may lack information as to the whereabouts of the sighting Roache provides the next wave of sightings with locations:

'in 1873 a sighting reported near what is now known as Talke Crossroads was followed by the disaster'. That disaster would be the Talk o' th' Hill disaster when 18 miners were killed as a result of a gas explosion.

He reported that further sightings took place at the areas known as Venice and Dunkirk which were closely followed by more mining accidents, the Bignall Hill disaster in 1874 when 18 miners lost their lives in a gas explosion and the Bunkers Hill 1875 mining disaster in which 43 miners were killed in an accident caused by blasting.

A nameless miner reported that while on his way home from the Diglake Colliery he was suddenly joined on his journey by a huge black fearsome hound of a type that he had never seen the like of, the dog vanished before his eyes. Following this there was of course the Diglake disaster in 1895 when 77 miners drowned as a result of flooding.

Roache summed it up thus: 'these disasters caused local people to associate the phantom's appearance with the disaster and to assume that when it was heard it was a sign of pending mining disaster.'

So far it does appear that sightings of the phantom dog had a good track record at advertising impending doom. But there were other sightings that appear to buck the trend.

When Roache was compiling his sightings of the black dog he interviewed a number of folk, one of these was the son of Mr MacGowan who had been the manager of the Harecastle colliery. He was said to have had a run in with the dog in 1886 when walking back from the colliery. He and his companion heard the sound of a large animal running towards them from behind, so convinced were they that there was something behind them that they parted and moved to either side of the path to allow the mystery animal to pass, however while they did feel the presence of the animal running past they saw nothing. For once there wasn't a mining disaster straight after this encounter, in fact the next accident was the Diglake one in 1895 and it would be stretching it to really connect the two events.

There have been more recent sightings, that is, within living memory depending on your age. Roache again provides the information this time thanks to Councillor Turner who had his sighting during World War Two along Boat House Road, Kidsgrove.

'I was going along boathouse road, a journey I made regularly over a period of many years, when a feeling of uneasiness came over me. The moon filtered through the branches of the overhanging trees casting an enchanting pattern on the roadway, and a great stillness settled over the area which itself was very unusual; persons knowing the area will confirm this. Suddenly a noise of wind swaying the trees burst out, but there was no wind, it died away and then I knew terror. I walked quickly and felt the road would never end. I had just reached the area occupied by Tarmac Limited, when I heard the sound of a large animal panting and running hard. I jumped in fright towards the stone wall, and the sound rushed past me and died in the distance taking the direction towards Bathpool... I hurried on and noticed as I came 'under the beeches' the night sounds were normal again. I have travelled the area many times since that awful night but I have never heard or felt anything unusual. I had never heard of the legend of the black dog until it was told to me at a later date. I have no doubt in my mind that I had encountered it, and can vouch for its truth.'

The next sighting that Roache recorded was from 1951, it was a little way from Kidsgrove at Mow Cop, a village that straddles the Cheshire/Staffordshire border. The sighting was made by a former policeman, Constable Stonehouse.

> 'I had been on duty many times in the Mow Cop area and being the Local Policeman I am familiar with it in every way. Early one morning I proceeded to the boundary of my area near the Mow Cop Inn, pushing my cycle. The scene was bright with the moon light and mounting my cycle I started to freewheel towards Mow Cop Church when I suddenly became aware of the sounds of an animal panting and running hard towards me, as it seemed from the lower slopes of Mow Cop, when I saw a huge dog jump from the hillside with coal black skin and red eyes, its tongue hanging out and in obvious signs of distress it ran into the churchyard. I had no fear at the time, but just a feeling of amazement, but to my horror, when I reached the churchyard, the iron gates were locked. I had just seen with my own eyes a large black dog go through them. It was only then I realised that what I had seen was of the supernatural, and that what I had in fact seen was the legendary black dog.'

Neither of these two sightings could be connected to any terrible event.

There have been further sightings reported since Roache compiled his list, the next set of reports were told to author and local historian Philip R Leese.

Mr C.W. Hopwood related the tale of how in 1939 he was living in Harriseahead and was courting a girl in nearby Kidsgrove. Walking home on a cold and frosty night around midnight he reached the top of Attwood Street when he had his run in with the black dog. The animal ran alongside him for a while, probably for a longer time than he would have liked as he described that he had been 'frightened to death'. However, soon enough the dog turned off and away from him and headed along the old dirt track to Newchapel. Once again there was no horrible outcome to the sighting other than a possible need for Mr Hopwood to change his trousers.

This was not the only tale that Leese was made aware of, in 1987 he received a letter from Nurse A. Wrigley. In the letter she recounted how in either 1945 or 1946 she had an encounter with the phantom dog. Having come to the district in 1943 she had never heard of the tales of the creature. As she didn't have a car at the time she would walk around the district, on her way back from Kidgrove she was taking a short cut across Engine Bank in the early hours, as it was summer it was already daylight. She sensed something was following her but she took no notice of it assuming it was a dog from a nearby farm, however when she got to the stile that led to Newcastle Road she turned to speak to it and was amazed to see a hound-like creature, light coloured and about the size of Great Dane which then darted through the hedge, but when she looked to follow it she couldn't see it, it appeared it had vanished.

At the time she put the experience down to a lack of sleep, but later that day when she was visiting one of her patients she recounted the tale to the woman who promptly told her husband 'You are not going to work tonight, Nurse has seen the boggart.'

The Mystery Animals of Staffordshire

The following day the nurse found out that there had been a fire at the place where the patient's husband was due to go to work. It appears that once again the dog was foretelling doom within the district, however it seemed to have lost its earlier body count of the heyday of unsafe mining.

Here see the mysterious dog being referred to as the boggart, the first time dog and boggart have had a connection since Hugh Bourne and his 1816 account.

We have already seen how the boggart managed to get a back history of being a murdered woman; the phantom dog itself managed to get its own back history of equally dubious origin. According to Norman Roache it all started with a pack of hounds owned by the Audley family who occupied the local castle and who controlled the local estate. Part of the rights they enjoyed was of course exclusive hunting rights to the deer that were to be found on their land, not everyone respected this arrangement which led to poaching taking place and to counter this free market enterprise the Audleys set loose their great hounds to roam the forest and keep intruders away from the deer. To stop these pesky hounds from ruining a day's hunting the poachers dug a great pit into which the lead hound fell into and having become stuck starved to death. Don't believe a word of that narrative, no, not one bit of it. Whatever the phantom dog of Kidsgrove is its shape- and size-changing nature would suggest it's not the leader of the pack.

As for sightings of the dog foretelling death it is again a bit subjective. While it certainly seemed to have timed its appearances to coincide with a number of mining disasters we have to remember it was of course at a time when the health and safety standards of mining in the country were shocking, if not just a shocking accident waiting to happen. It would appear therefore that sightings of the dog regularly happened and that occasionally they happened to coincide with a mining disaster, as has been seen there are plenty of times when the dog appeared with no bad news attached.

After many years of regular appearances the phantom hound of Kidsgrove seems to have vanished. There could be many reasons for its no show in recent years, for a start there are no longer mines in the area that need a phantom hound to appear to announce an accident waiting to happen. And it's not only the mines that have disappeared, so have many of the fields that the hound once roamed, these now form parts of a number of housing estates. Perhaps we should also consider that most of the sightings were made by people walking whereas today many of those journeys would be made by car and the chances of hearing a heavy breathing hound while sitting in your car with the radio on are slim. Before you start to think that that's the end of the phantom dog its worth considering that on many of the occasions it was reported it was only felt, as of course it was an invisible dog. And just how do you go about spotting one of those?

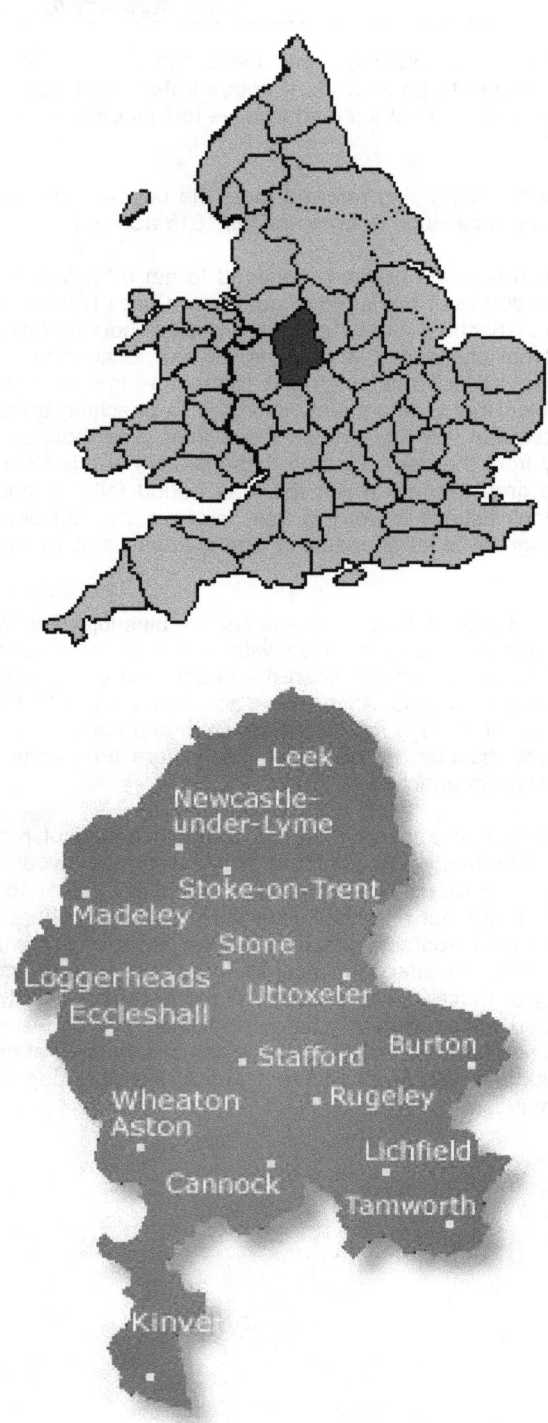

Chapter II
Water-Maidens
'...It was a mermaid, the same as you read of in the papers...'

While the notion that the county of Staffordshire might actually be the home of living, breathing, flesh-and-blood mermaids will inevitably be greeted by many with the rolling of eyes and hoots of derision, it is an undeniable and astonishing fact that such beliefs persisted for centuries – and, in those parts of Staffordshire where the many and varied traditions and superstitions of times-past can still be found lurking, that belief actually quietly continues.

The word 'mermaid' is derived from a combination of 'mere', the Old-English word for sea, and 'maid': a woman. According to old sea-faring legends, mermaids would often deliberately sing to sailors to try and enchant them; with the secret and malevolent intent of distracting them from their work and causing their ships to run disastrously aground. Other ancient tales tell of mermaids inadvertently squeezing the last breaths out of drowning men while attempting to rescue them. They are also said to particularly enjoy taking humans to their underwater lairs. In Hans Christian Andersen's *The Little Mermaid*, for example, it is said that mermaids often forget that humans cannot breathe underwater; while other legends suggest the sinister she-creatures deliberately drown men - out of sheer, venomous spite, no less.

The fabled sirens of Greek mythology are sometimes portrayed in later folklore as being mermaid-like in nature and appearance; in fact, some languages use the same word for both bird and fish creatures, such as the Maltese word, Sirena. Other related types of mythical or legendary creatures include water-nymphs and selkies: animals that can allegedly transform themselves from seals into human-beings - and vice-versa, too.

Mermaids were noted in British folklore as being distinctly unlucky omens – occasionally foretelling disaster and sometimes even maliciously provoking it, too. As evidence of this, several variants of the ballad, *Sir Patrick Spens*, depict a mermaid speaking to the doomed ships. In some, she tells the crews they will never see land again, and in others, she claims they are near shore, which the men are wise and astute enough to know means that deep, malevolent deception is at work.

The ballad itself is one chiefly of Scottish origins, and may possibly refer to an actual event: namely, the bringing home of the Scottish queen Margaret, Maid of Norway across the North

Sea in 1290. There is, however, some speculation that the ballad may actually relate to a voyage by the princess's mother in 1281. But, regardless of the specific truth behind the ballad itself, its words are prime evidence of both the knowledge and the deep fear of mermaids that has existed in the British Isles for an untold number of centuries.

Before addressing the matter of the mermaids of Staffordshire, however, it is worth noting that in Shropshire – a county which borders directly upon Staffordshire and which shares many of Staffordshire's ancient folkloric tales and legends – there are a number of old stories of the very mermaid-like variety. One such account tells of a deadly mermaid inhabiting a small pool in the pleasant little village of Childs Ercall. In 1893, the writer Robert Charles Hope described the story as follows:

> '...there was a mermaid seen there once. It was a good while ago, before my time. I dare say it might be a hundred years ago. There were two men going to work early one morning, and they had got as far as the side of the pond in [a] field, and they saw something on the top of the water which scared them not a little. They thought it was going to take them straight off to the Old Lad himself! I can't say exactly what it was like, I wasn't there, you know; but it was a mermaid, the same as you read of in the papers. The fellows had almost run away at first, they were so frightened, but as soon as the mermaid had spoken to them, they thought no more of that. Her voice was so sweet and pleasant, that they fell in love with her there and then, both of them. Well, she told them there was a treasure hidden at the bottom of the pond - lumps of gold, and no one knows what. And she would give them as much as ever they liked if they would come to her in the water and take it out of her hands. So they went in, though it was almost up to their chins, and she dived into the water and brought up a lump of gold almost as big as a man's head. And the men were just going to take it, when one of them said: "Eh!" he said (and swore, you know), "if this isn't a bit of luck!" And, my word, if the mermaid didn't take it away from them again, and gave a scream, and dived down into the pond, and they saw no more of her, and got none of her gold. And nobody has ever seen her since then. No doubt the story once ran that the oath which scared the uncanny creature involved the mention of the Holy Name.'

Of course, this case – as I have noted above – was not Staffordshire-based; however, I make specific mention of it to point out that within central England, in centuries-past, belief in the mermaid was without doubt widespread. And with that said: now onto neighbouring Staffordshire.

The village of Thorncliffe, near Leek, on the Staffordshire Moorlands, has a very memorable tale attached to it of a mermaid that can supposedly be seen at the witching-hour, at the appropriately named Mermaid's Pool. For those who may be unacquainted with the Moorlands, they are typified by forests, lakes, rolling hills and crags and have the distinction of being the home of Flash: the highest village in the British Isles, which stands at 1,518 feet above sea-level.

But back to the tale of the mermaid: those that get too close to the seemingly beautiful, flirty creature, as she tantalisingly and teasingly combs her long and flowing locks under a starry, moonlit sky, are destined to be dragged into the waters of the pool by what is in reality a malevolent, and utterly deadly, she-devil of a creature, tellers of the tale say.

Reputedly, the legend dates back to around the tenth century when a young girl (who, so the story goes, may have been a witch and one who was very well-practiced in the world of the black-arts) was pursued and persecuted by a local man, who duly threw the girl to her death in the waters of Mermaid's Pool. She, in turn, proceeded to scream absolute bloody vengeance upon her persecutor before she finally disappeared under the water and was duly drowned. And, sure enough, the man's body was shortly thereafter found dead in Mermaid's Pool – his face violently torn to pieces, as if by monstrous and vicious talons.

One person who has dug deeply into this particular story, and who can shed yet further significant light on it and fill in some of the blanks, is Lisa Dowley, who says of her own investigations, memories and experiences:

> 'As small children during the seventies, my sister and I were often taken out for lunch on Sunday afternoons, as our parents owned a public-house. One of our favourite jaunts was a drive through the Staffordshire moorlands, stopping at a quaint, but remote inn for lunch. During our many journeys our father would tell us many a tall tale, and on an odd occasion we were lucky enough to glimpse the wallabies on the moors [Author's note: about which, much more indeed in a later chapter]. However, the story that was always told to us, and which remains in my mind the most, is the tale of the mermaid.
>
> '"And this is the place where they deal with naughty girls," our father would say to us, in a stern, fatherly manner as we approached the top of a steep incline. My sister and I would turn to each other with looks of grave concern, as our behaviour between us on most days was less than satisfactory to our parents, or anybody else for that matter.
>
> '"Yes, it's true," he continued, "many, many years ago a woman was drowned her for being naughty and wicked. When she died", he continued, his voice getting slower, sterner and deeper, "she changed into a mermaid and now waits for other naughty girls to join her!" He never did divulge to us the reason for her unpleasant death – not that we ever asked.
>
> 'Now, on the face of it this may not read as too scary, but when you are six and seven years of age respectively, it tends to make a lasting impression. Even in the hot summers of the 70s, my sister and I were loathed to get out of the car and admire the view with our parents; the drop off the side of the road to us looked at least a hundred feet. We always preferred to stay in the car; our parents probably viewed this as a success on the good behaviour and peace-and-quiet front.

'Even on warm summer afternoons, the Blakemere at Morridge – as I recall – has retained an eerie and slightly oppressive sense about it. However, its expanding moorlands and remote bleakness still manages to be invitingly beautiful, while retaining that eternal mystical feel that totally wraps itself around you, even to this day.

'Many years later, after returning to my home county of Staffordshire, the stories that our father told were still quite vivid and swirling around my head, so I decided to find out if there was any truth – once and for all – in the tales that had been told to us. Curiosity has always been at the heart of my nature, and as I had always found it odd, even as a small child, that there should be a pool so far inland that was home to a mermaid, at one of the highest points, if not the highest place in England.

'Upon investigation, there seem to be many mysterious, murderous and mythical tales which may or not be merely folklore, that surround this bleak and remote part of the Staffordshire moorlands. Written origins of this particular mermaid folklore tale can be traced back to some 1,000 years ago.

'The story transpires that this particular mermaid was once a maiden of fair beauty, and it came to pass – for reasons that are unclear – that she was persecuted, and accused of various crimes by a gentleman, named Joshua Linnet. It is not clear whether these accusations included being a witch, or whether he may have had his amorous advances rejected.

'The said Mr. Linnet had this woman bound up, and thrown into the bottomless Blakemere pool. As she fought for breath and life, the woman screamed vengeance on her accuser Joshua Linnet, and that her spirit would haunt the pool from that moment hence, and swore that one day she would drag her accuser and executioner deep down beneath the dark depths of the Blakemere to his own death.

'It is a recorded fact that three days later, Joshua Linnet was found face down, dead in the Blakemere pool. When his body was dragged out and turned over by the locals, to their horror, what greeted them was that what was once his face was now nothing more than tattered shreds of skin, the injuries seemingly caused by sharp claws or talons.

"The mermaid of Blakemere is no bringer of good fortunes; rather she will entice any passer-by and take then down beneath the still black waters to certain death. This is reinforced on the wall of the pub of the same name (*The Mermaid Inn*, formally Blakemere House) which is situated just below the Blakemere pool. On the wall it is written: "She calls on you to greet her, combing her dripping crown, and if you go to greet her, she up and drags you down."

'The Blakemere, for many centuries, continued to inspire speculation and instill fear among locals, and the belief in the mermaid persisted well into the nineteenth century. However, in the seventeenth century, a sceptic called Dr. Robert Plot, a historian and naturalist of the time, was told of the legend of Blakemere and that no animal would drink in its bottomless waters or any bird fly over it. All of which he disputed in his 1686 publication *Natural History of Staffordshire*.

'Towards the end of the nineteenth century work was began in an attempt to drain the Blakemere. However, as the workmen attempted to drain the pool, they reported that the mermaid appeared to them and warned that if the waters of the Blakemere were removed, the whole town of Leek would be drowned. And so the men fled from the Blakemere on Morridge refusing to go back to complete the work, and so the pool remains to this day untouched and un-drained.

'There are a number of mermaid traditions and legends that are associated with numerous pools within the locality of the Staffordshire moorland borders, and evidence to suggest their origins may well have been rooted in pre-Roman Celtic traditions. However, if you wish to pay the mermaid of Blakemere a visit, it is said the best time to chance seeing her is around the midnight hour. But take care, as the Blakemere pool now resides on Ministry of Defence land, so be diligent in your steps so as not to tread on any soldiers on night manoeuvres! Or you may prefer to view the Blakemere pool from the roadside from within the safety of your car.'

There ends Lisa Dowley's thoughtful and thought-provoking account. We are not yet done with the Staffordshire mermaids, however; not by a long-shot.

What is without doubt the absolute oddest, and potentially the most disturbing, story that may be related to the theme of this particular chapter, comes from a family who, in an October 2000 interview with me, maintained that while driving near the site of the 1643 Battle of Hopton Heath on a cold winter morning in 1979, they came across a shocking sight: namely, the badly damaged body of a very strange-looking young woman sprawled at the side of the road 'that looked like a hit-and-run'. The woman, says the resolutely-anonymous family, was naked, had a head of long, dark hair, a mouth full of elongated teeth, and a pair of legs that were fused together below the knees, 'like a seal'!

The story gets even more controversial by the fact, or rather the claim it should be clarified, that the family elected not to inform the authorities of what they had found, and quickly continued on their journey. To this day, their action – or, far more accurately, their utter *lack* of action – hangs over them like a veritable sword of Damocles; or so they collectively assert, at least.

Of course, it does not even need to be said that this particular story is outrageous in the extreme, and numerous questions remain wholly unanswered. For example, even if the tale has a semblance of truth to it, and the family in question failed to inform the emergency services of their surreal and grim discovery, then why did no-one else stumble upon the body and take the initiative instead? Also: if their tale is accurate, then where, precisely, did the body end up?

While these very valid questions are all strongly suggestive of nothing more than a somewhat tasteless hoax, one might be tempted to ask: what would prompt the family to create such an audacious story, when neither publicity nor payment played any sort of role in their decision to relate the story to me?

Unless, or until, someone else comes along to offer some type of verification or vindication, or who can ensure that we fully dismiss it out of hand, this strange account will continue to remain in limbo as one of those curious, rogue cases that surfaces now and again. I am very sorely tempted to dismiss it as a mere hoax and absolutely nothing else at all; but then again, who can truly say for certain what monstrosities and malformations nature may throw up from time to time?

The case is also made all the more intriguing for another reason: as long-term students of anomalous phenomena will be acutely aware, in locales where one mystery can be found, so – very often – can others, too. And that is most certainly the case with Hopton Heath.

A key event of the First English Civil War, the aforementioned Battle of Hopton Heath was fought on Sunday, March 19, 1643 between Parliamentarian and Royalist forces. The battle ended at night-fall, with the actual victory and outcome still remaining matters of very much personal opinion to this day: the Royalists, for example, had succeeded in capturing eight enemy guns; while the Parliamentarians believed that their successful killing of the enemy commander, the Earl of Northampton, was of equal – if not even greater – significance.

More than 300 years later, one night in the winter of 1974, John 'Davy' Davis, who was aged thirty-six at the time, and living and working as a painter-and-decorator in the city of Lichfield, Staffordshire, was driving near the site of the battle when he began to feel distinctly unwell: a tightness developed in his chest, an odd feeling of light-headedness came over him, and his 'left ear hurt and felt hot'.

Pulling quickly over to the side of the road, Davis was amazed to see the night sky suddenly transform into a sudden cascade of daylight, while the road that should have been in front of him no longer existed: instead, it had been replaced by a mass of fields, heath and sparse trees. And before his unbelieving and astonished eyes, countless soldiers adorned in Civil War period clothing fought each other savagely.

Notably, Davis said that although at one point he was 'nearly bloody surrounded' by the soldiers, it seemed as if they could neither see him nor his vehicle. To a degree, at least, this afforded Davis some much welcome relief, as he was practically frozen to the spot; unable to drive away, as much as he had surely wanted to. As it transpired, Davis didn't need to go anywhere at all: only a few seconds later, the bizarre scene suddenly vanished, and Davis found himself sitting at the side of the road, with his car squashed hard against a line of hedge.

The Mystery Animals of Staffordshire

All had returned to normal.

Did John Davis *really* experience some form of weird time-slip? Or was the brief affair all the result of some strange, wholly internal hallucination brought on by a seizure? To his credit, Davis is quite open to the latter explanation, but remains puzzled by the sheer vivid and graphic nature of the whole experience.

Of course, this particular event of 1974 has nothing, directly at least, to do with the affair of the mermaid alleged to have been stumbled upon in the same area five years later. But, it surely demonstrates that the location of – and the area that surrounds – the Battle of Hopton Heath is a highly unusual one, indeed. And things are about to get weirder, still, as we dig into the story of the malevolent and deadly water-maiden of Aqualate Mere.

Situated barely a stone's throw from the Shropshire town of Newport and just over the border into rural Staffordshire, Aqualate Mere – at 1.5 kilometres long and 0.5 kilometres wide - is the largest natural lake in the Midlands; yet it is very shallow, extending down to little more than a uniform three-feet. Well hidden on a private estate by low-lying woodlands that are themselves dominated by alder and willow, fen meadow and wet pasture land, Aqualate Mere is home to a rich variety of wildlife, including the buzzard, the barn-owl, the mallard, the teal, the fox, the polecat, the otter, the mink, both pike and bream; and – notably and somewhat surprisingly, too - a thriving, healthy herd of Exmoor ponies. But it may be home to far stranger beasts, too.

Legend has it that one day many years ago, when the Mere was being cleaned, a mermaid violently rose out of the water – quite naturally scaring the living daylights out of the workmen – while simultaneously making shrieking, disturbing and damning threats to utterly destroy the town of Newport if any attempt was ever made to empty Aqualate Mere of its precious waters. Very wisely, perhaps, the Mere was not – and, to date, never has been - drained.

Of course, this story is practically identical to that related earlier by Lisa Dowley and which concerns the Blakemere Pool. Undoubtedly, this suggests a likely common point of origin for both accounts – and, therefore, a distinct possibility that both cases are strictly folkloric in origin and nature, and nothing else. Nevertheless, I suggest that one and all should still be very careful when negotiating the many mysterious bodies of water that can be found throughout all of Staffordshire. One never knows just what might be lurking beneath the surface and within their murky, impenetrable depths...

Glen Vaudrey
Me again. So you've read about the mermaids that Nick found, now it's my turn to introduce you to another type of soggy lass.

While she isn't a mermaid there is another woman of the water to be aware of in Staffordshire and that is Jenny Greenteeth, no not someone in desperate need of a dentist but a strange folkloric health and safety creature. Why a health and safety creature? Well there are numerous folkloric creatures that are said to be nothing more than a tale to tell to young children in order to scare them away from water and the risk of drowning. However just supposing this isn't the case and there really is a murderous water dwelling creature out there I'd better give a description.

The beast known as Jenny Greenteeth possessed a pale green skin, very long lank green hair, and if that wasn't bad enough there was also the long green fingers and horribly green nails and, as the name suggests, green teeth. Let's face it she was no looker, so it's no wonder she hid at the bottom of still pools of water. Of course some folk claim that no part of her was ever spotted above the water as she always stayed hidden below the weeds covering the surface.

She was equally at home in natural or manmade water features be it pond or canal, just how she arrived in a new pond is never really explained but once the weed appeared you could be sure that she would be lurking beneath it. Somewhere below a layer of green algae or duckweed she would lay in wait and it was said if you came too close to the edge then a pair of arms would shoot out and grab you before dragging you to a watery grave only for the duckweed to close up over you so that no one would ever know your fate.

Staffordshire wasn't the only county where Jenny Greenteeth could be found she also haunts the pools of Cheshire but that's for a different book.

Chapter III
Out-of-Place Cats

'...It is impossible to say categorically that no big cats are living wild in Britain...'

During the early part of 1998, the British Government's House of Commons held a fascinating and arguably near-unique debate on the existence – or otherwise – of a particular breed of mystery animal that is widely rumoured, and even accepted by many, to inhabit the confines of the British Isles: the so-called Alien Big Cats, or ABCs, as they have become infamously known.

It scarcely needs mentioning (but I will do so, anyway!) that Britain is not home to an indigenous species of large cat. Nevertheless, for decades amazing stories have circulated from all across the nation of sightings of large, predatory cats that savagely feed on both livestock and wild animals and that terrify, intrigue and amaze the local populace in the process. And, of course, the media loves them, one and all!

As history has demonstrated, there now exists a very large and credible body of data in support of the notion that the British Isles do have within their midst a healthy and thriving population of presently unidentified large cats – such as the infamous Beast of Bodmin and the Beast of Exmoor that so hysterically dominated the nation's newspapers back in the early-to-mid 1980s. And, I admit I'm very pleased to say, much of that credible evidence comes from the county of Staffordshire. Before addressing the big cats of this fair and fine location, however, let us turn our attentions to what can be – and indeed has been - determined at an official level about the puzzle.

Documentation that was generated as a result of the February 2, 1998 debate on the controversy in the House of Commons began with a statement from Mr. Keith Simpson, the Member of Parliament for mid-Norfolk: 'Over the past twenty years, there has been a steady increase in the number of sightings of big cats in many parts of the United Kingdom. These are often described as pumas, leopards or panthers. A survey carried out in 1996 claimed sightings of big cats in 34 English counties.'

Many of the sightings, Simpson continued to the House, had been reported in his constituency by people out walking their pet-dogs or driving down old country roads, very often at dawn or dusk. Frequently the description given fitted perfectly that of a puma or a leopard. Simpson also added that in a number of incidents it had been claimed that ewes, lambs, and even horses had been attacked – and in some cases killed – by the marauding beasts.

Simpson elaborated yet further:

> 'A number of distinguished wildlife experts have suggested that some pumas or leopards could have been released into the countryside when the Dangerous Wild Animals Act 1976 made it illegal to own such animals without a licence. They would have been able to roam over a wide area of countryside, live off wild or domestic animals and possibly breed. So what is to be done?'

In answer to that question, Simpson had a few ideas of his own:

> 'I should like to suggest two positive measures for the Minister to consider. At national and local levels, it is logical that the Ministry of Agriculture, Fisheries and Food should be the lead Government Department for coordinating the monitoring of evidence concerning big cats.'

In response, Elliot Morley, at the time the Parliamentary Secretary to the Ministry of Agriculture, Fisheries and Food, admitted that there was a valid issue that did need addressing. He said:

> 'The Ministry's main responsibility on big cats is confined to whether the presence of a big cat poses a threat to the safety of livestock. The Ministry is aware that a total of 16 big cats have escaped into the wild since 1977. They include lions, tigers, leopards, jaguars and pumas, but all but two animals were at large for only one day.'

Morley expanded still further:

> 'Because there is a risk that big cats can escape into the wild and because of the threat that such animals could pose to livestock, the Ministry investigates each report in which it is alleged that livestock have been attacked. Reports to the Ministry are usually made by the farmers whose animals have been attacked. In addition, the Ministry takes note of articles in the press describing big cat incidents and will consider them if there is evidence that livestock are at risk.'

On receipt of a report of a big cat sighting, explained Morley to the House, the Ministry would ask the Farming and Rural Conservation Agency – the Ministry's wildlife advisers – to contact the person who reported the encounter, as he explained: 'The FRCA will discuss the situation with the farmer and seek to establish whether the sighting is genuine and whether any evidence can be evaluated. It will follow up all cases where there is evidence of a big cat that can be corroborated and all cases where it is alleged that livestock are being taken.

> 'The FRCA will consider all forms of evidence, including photographs given to it by members of the public and farmers, plaster casts of paw prints and video footage. In addition, it will carry out field investigations of carcasses of alleged kills for field

signs of the animals responsible.'

In conclusion, Morley stated:

> 'It is impossible to say categorically that no big cats are living wild in Britain, so it is only right and proper that the Ministry should continue to investigate serious claims of their existence – but only when there is a threat to livestock and when there is clear evidence that can be validated. I am afraid that, until we obtain stronger evidence, the reports of big cats are still in the category of mythical creatures.'

Thanks to the Freedom of Information Act, we now have that 'stronger evidence'. Replying in 2006 to a FOIA request from a member of the public with an interest in big cat sightings seen in the county of Hampshire between 1995 and 2005, the county's Police Force released secret files that stated:

> 'Hampshire's Constabulary's Air Support Unit has been deployed to assist with the following reports: January 1995 – Black Panther like animal seen in Eastleigh. Two likely heat sources found by the aircraft, but nothing found by ground troops. March 1995 – Black Puma like animal seen in Winchester. One heat source found that could not be classified by the aircraft crew, kept running off from searching officers, search eventually abandoned.'

Notably, when a similar FOIA request was filed with Sussex Police in late 2005, documentation was made available to the requester that read as follows: 'Firearms officers have been deployed in response to such a report on one occasion, on 22 July 2004 – sighting by a member of the public in Seaford. The area was searched, but no trace was found of such an animal.'

The story is far more spectacular on the east coast of England, however. In 1991, documents show, a lynx – that the Department for Environment, Food and Rural Affairs believed may have escaped from a zoo; although this was never actually proven – was shot dead near Great Witchingham, Norfolk, by a man who then placed the body in his freezer before selling it to a local collector who decided to have the creature stuffed.

It transpires that an extensive dossier on the affair was opened by local police that – as with the above-reports on other exotic felines prowling the British countryside – would have remained under lock-and-key were it not for the useful provisions of the Government's Freedom of Information Act.

It all began when police officers were investigating a gamekeeper who, it was suspected, may have been responsible for the deaths of a number of birds of prey that had been living within the area. The officer that interviewed the man in question wrote in his now-declassified official report:

> 'At the start of the search in an outhouse, which contained a large chest freezer, I

asked him what he had in the freezer, and he replied: 'Oh, only some pigeons and a lynx." On opening the freezer there was a large lynx lying stretched out in the freezer on top of a load of pigeons! He had shot this when he saw it chasing his gun dog.'

Britain's big cats, it seems, are no longer the myth that many want them to be or believe them to be – and the government knows it full well, too. As do the many and varied people of Staffordshire, as now becomes more than graphically apparent...

Chapter IV
Creatures on the Loose
'...I thought he looked a bit big for a tabby...'

One of the earliest, credible reports of a big cat encounter in Staffordshire took place one morning in the latter part of the 1980s. The witness was a man who now lives in the Staffordshire town of Cheslyn Hay and who contacted me by telephone after reading a June 2006 interview with me that the *Chase Post* newspaper – that covers the town of Cannock and the surrounding Cannock Chase – had conducted on the subject of the myriad monster-driven mysteries of the area.

Very careful to not reveal his true identity – even his telephone number was deliberately blocked – the man somewhat guardedly and hesitantly advised me that he had been driving to work on his motorbike along the Penkridge Road out of Cannock at around 7.30 a.m. on the day in question; as he did on each and every weekday morning. This particular journey, however, would be completely unlike any other that he had experienced; either before or since, as it transpires. As he approached the town's old railway bridge, the man saw a strange-looking creature directly in front of him and at a distance of approximately several hundred feet: around twenty inches in height and possessing a long tail, it was undoubtedly a large, black cat. And, we aren't talking about your average household puss either; not at all.

As the man sped directly towards the cat, he could only watch with a mixture of shock and relief as it suddenly shot across the road, quickly headed down the embankment and utterly vanished from sight. The man stressed that he never saw the creature again, but was most assuredly convinced that what he had seen was indeed a large cat that very much resembled a puma.

Having quickly taken notes as the man carefully related the facts, I asked him if it would be okay to quote him within the pages of this book. 'Well...alright, I suppose so,' he said, somewhat reluctantly, before terminating the call – and somewhat abruptly at that too, I thought to myself. His identity has always remained elusive to me. His account, however, remains forever memorable.

Staffordshire-based big cat encounters in the 1990s – or, perhaps more correctly, *reported* encounters - were relatively few and far between. In other words, we can never be quite sure how many people remained silent with respect to their sightings during that decade, carefully preferring to avoid overwhelming media publicity and potential ridicule.

One encounter from the 1990s – specifically from September 1996 - that *does* deserve a place within the pages of this book, however, occurred on the fringes of the town of Penkridge. The witness was a local postman, who, while delivering mail on his regular early morning route, almost literally walked slap-bang into what he was able to determine – after studying a book on big-cats – was without any shadow of doubt a fully-grown lynx.

The postman, in surprisingly unexcited tones, stated that:

> 'I thought he looked a bit big for a tabby, but I went on walking. But when he got closer, he just looked up out the corner of his eye at me, and grumbled and growled at me a bit as he walked past, and carried on walking up the path, jumped on the wall of the bloke's next-door and went on his way into his back garden. After it was over, I felt a bit like: "Blow me: that was like one of those you see in the newspapers." He was a big lad; but not a lion or a tiger, obviously. And it was all over, anyway, before I could have got scared and actually knew what he was.'

As a new century dawned, it seemed that the big cats of Staffordshire were determined to make their presence known to just about one and all, and whenever and wherever conceivably possible, too.

In January 2000, a Cannock-based journalist named Rory McBride made it his very own job to seek out the startling truth for himself. An article written for the *Chase Post* newspaper and titled 'Here, Puss, Puss, Puss!', revealed that McBride's quest had been prompted by the dedicated research of a man named Robin Roberts, a head-keeper at Drayton Manor Zoo, who had tirelessly devoted much of his free time over the course of the previous five years to uncovering the stark and sensational truth behind the controversy of Staffordshire's big cats.

Roberts told McBride that:

> 'There have been a significant number [sic] of sightings of large black cats on Cannock Chase, which are probably black panthers or leopards. The habitat could support a population of big cats. There is enough space for them to live and more than enough food.'

Roberts elaborated further:

> 'I think there must be a breeding population because our oldest lions live at best for twenty or so years in captivity where they are treated by vets. Cats released by collectors in the 1970s must be breeding because big cats do not live that long in the wild.'
>
> 'If you are trying to track down the beast, then I wish you good luck,' Roberts advised the intrepid *Chase Post* reporter. Nevertheless, he had both cautionary

and ominous words to add: 'It could be within a few feet of you, but concealed in the shadows and undergrowth. You will not see it unless it wants you to. It will live in the quietest part of the wood and travel around long disused trails and train tracks. There may also be signs of a scratching post. People should not be alarmed; the cats would not attack a human unless they were cornered. The last thing we need is idiots with guns trying to hunt them.'

Spurred on by Robin Roberts' investigations, Rory McBride recorded in his write-up that:

'I set out on an expedition to Cannock Chase, hindered only by a swirling mist. I searched caves, crannies and caverns with military precision. Scanning the horizon with powerful binoculars, I sensed a chilling eye boring into my back, but turned around to see nothing but shadows. Tracks ignited enthusiasm, but led only to dead ends. The search goes on...'

Barely a month after journalist Rory McBride was hot on the trail of the big cats of the Cannock Chase, the report of Robert and Jean Beeston of the nearby town of Heath Hayes was made public. The sighting had occurred on a balmy summer's evening during the previous August as the pair drove past the German Cemetery – the site of numerous close encounters of an extremely strange and animalistic kind over the years.

Situated near to Broadhurst Green, the cemetery is a memorial to no less than 5,000 German servicemen, whose graves are marked by headstones constructed out of Belgian granite and set in plots of heather. A tribute to the spirit of cooperation that now exists between the Commonwealth War Graves Commission and the Volksbund Deutsche Kriegsgraberfursorge – the German War Graves Commission - the cemetery had its origins in 1959, the year in which Germany made the first, initial approaches with a view to finding a site on the Chase near the existing Commonwealth Cemetery that contains the graves of 388 men from both World Wars One and Two, including 287 German soldiers.

In March 1962, the County Council made a gift of the land to the German Government, with the design of the cemetery and its surrounding buildings placed in the expert hands of Professor Diez Brandi of Gottingen, Germany, and Harold Doffman and Peter Leach, who, at the time, were partners in a Stafford firm of architects.

As a result of the construction of the cemetery, between 1964 and 1966 the bodies of numerous German servicemen, sailors buried at seaports around the British coast, airmen shot down inland, and soldiers – most of whom were prisoners-of-war buried in churchyards around the country – were transferred to the cemetery, which today can boast of thousands of visitors per year, including some of a very odd and beastly type.

In the case of Robert and Jean Beeston, for example, who saw a big cat near the cemetery in 2000, although the couple only held the creature in their sights for mere moments, Robert Beeston was adamant that:

'It was a big, black panther. It was massive; with one leap it cleared the road. We weren't afraid; it was just an incredible sight.'

'Stop the Beast!' That was the wild headline that practically screamed in wholly hysterical fashion from the pages of the *Chase Post* precisely a week after the Beeston's sighting surfaced. It had been prompted by the words of Heath Hayes man Bill Moseley who 'vowed never to go walking on the Chase again' unless an investigation into the activities of 'the beast' was undertaken.

That Moseley had seen one of the elusive animals for himself, around 10.00 p.m. near the *White House Hotel* on the Penkridge Road in September 1999, made his comments somewhat understandable, it must be admitted. Rather ironically, the *White House Hotel* itself has a connection to the world of all-things-unexplained: it was at the hotel, in September 1996, that the now-sadly-defunct *Staffordshire UFO Group* held its first, annual conference on matters of an extraterrestrial, bug-eyed and conspiratorial nature.

Moseley elaborated:

> 'I have no doubt as to what I saw. The animal was a panther, about six-foot in length with a long tail and very powerfully built. I was shocked to see how really big it was. I stopped the car and turned to my daughter and asked her what she thought she had seen. She also had no doubt it was a panther.'

And Moseley had much more to add, too:

> 'We informed the police when we got back but I don't think they took us seriously. That's the trouble: people aren't taking it seriously. Does it need this animal to seriously injure or kill someone before people take notice?'

Admitting that prior to his own encounter he was 'sceptical' of such reports and that he 'needed to see the animal before I believed', Moseley concluded with words that no doubt struck a deep and echoing chord with many concerned readers of the area's most popular newspaper:

> 'I have two younger children who have played close to where the animal ran out. Sooner or later there is going to be an accident; you simply wouldn't stand a chance with that thing. Neither myself nor my kids will be setting foot on the Chase before something is done.'

From that very moment onwards and in the eyes and minds of many at least, the Cannock Chase big cats were no longer merely just wild cats amiably roaming the woods and feasting on the occasional rabbit, deer or squirrel. Rather, to them, the creatures were now nothing less than potential man-eaters; and complete and utter tragedy was possibly looming just around the next fir-tree-shrouded corner.

Almost certainly as a result of the publicity and fanfare that was afforded the reports by local media outlets in January and February 2000, an absolute wealth of new data surfaced in March of that year, including the report of Keith Ball, who had seen a big cat on Cocksparrow Lane as he headed home after visiting a relative at the nearby Stafford Hospital.

Ball said that he traveled the same lane every night, but on the evening at issue he had just gone under a road bridge when something very strange was caught in the glare of his headlights: 'In the distance I could see two points of light coming towards me. From the way the light was moving I thought it was someone running with a torch.'

Needless to say, the two points did not emanate from a torch: rather, they were the illuminated eyes of a truly immense cat. Ball added: 'It was on the other side of the road and it just ran past me, heading towards the *Barn* restaurant.'

Keith Ball's account was mirrored by that of a 68-year-old Hednesford man who declined to be named, but who had seen just such a creature in the area on no less than two occasions. He carefully recalled that: 'One Monday lunchtime a few weeks ago I was driving home from *Ingestre Park Golf Club* when this cat jumped over the hedge, then pounced over the ditch. This cat was about 18 to 20 inches high with a long, thick black tail.'

The man then proceeded to inform staff at the golf club on the following day, who responded in a positively and surprisingly underwhelming fashion: 'I assured the woman I had not been drinking but she didn't seem bothered.'

It was two or three weeks later that the man saw the animal – or, at the very least, a similar one – once again. The time was roughly 6.30 p.m. and it was already getting dark when, while heading for the Golf Club and having just passed the Forestry Commission offices, his headlights picked up something at a distance of about fifty or so yards: 'I saw a pair of green eyes in the hedge. I tooted my horn at it and it ran off.'

Somewhat oddly and intriguingly, he added: 'Since then I have heard a lot of people are walking with big torches around there. Last Monday I saw about ten people who appeared to be searching for something.' Precisely who these torch-wielding and mysterious characters were, and: what they were doing in the woods of the Cannock Chase remains a distinct and thought-provoking puzzle to this very day.

And it wasn't just the general public who were reporting creepy, big cat activity in the area either: Senior-Ranger Mark Payne revealed that many of his colleagues were privately inclined to think that there was possibly 'an unknown predator out on the Chase'. Payne stressed that: 'I am very open-minded and I would say we're all of the same opinion. I wouldn't say there isn't a big cat out there because there have been a lot of releases of panther type of creatures.'

But Payne had more to impart than that; much more, in fact. Several years before, near Spring Slade Lodge, said Payne: 'I found a print as big as my fist by a dead deer. It looked like the print of a cat. But what really impressed me was the way the deer had been eaten – the outer skin had been peeled back which is not what you'd expect from a badger or a fox.'

Emphasising the sheer quality of witnesses that were beginning to come forward, Payne added: 'Responsible people are reporting this animal – we even had someone from the rural crime squad who told us he'd seen it.'

In the early part of March 2000, thirty-two-year-old Hednesford-based big cat enthusiast Alan Harrison revealed the following: 'The other weekend I was walking with a friend in Brindley Heath near Pars Warren when we found two massive footprints. We also noticed there were no claw prints showing – typical of cats rather than dogs – so we took plaster casts of them.' Indeed, an impressive photograph of Harrison and his intriguing piece of evidence took pride of place within the pages of the *Chase Post*.

Harrison had much bigger and far more ambitious plans, too – and they extended way beyond merely taking plaster casts, as he revealed:

> 'I have bought a CCTV style camera with a wide angle lens to set up in my Landrover which will pick up infra-red images of anything that's there. I'm also hoping to get permission from the Forestry Commission to get vehicle access to parts of the Chase. If I can plot on a map where people have seen the panther recently it should improve the chances of tracking it down.'

Unlike many frightened locals who, from the safety and comfort of their homes, were practically baying for the blood of the beast or beasts, Harrison had a refreshingly different attitude towards resolving the matter once and for all: 'I would like the panther to be caught, tagged, [and] then released so people can track where he is.'

Midway through 2000, *Chase Post* editor Mike Lockley asked me and Irene Bott - the then-President of the *Staffordshire UFO Group* - if we would be interested in writing a regular column for the newspaper on the subject of paranormal, unexplained and mysterious activity in the vicinity of the Cannock Chase. Of course, we quickly said 'yes', we most certainly *would* be interested. In addition to giving us the absolute perfect platform for highlighting the results of our combined research, our column – christened, somewhat predictably, as *The C-Files* and that saw me playing the role of Mulder to Irene's Scully – also ensured us a huge amount of feedback from readers with their own accounts of the very strange kind to relate.

One such reader was a man named Ryan Jarvis who, in August 2000, had been driving to work around 5.00 a.m. when, as he drove past the *Wych Elm* pub, saw ambling across the car-park of the nearby *Grosvenor Working Men's Club* a 'Rottweiler-sized' big cat that was of a 'tortoise-shell-brown colour'. Not quite believing his own eyes, Jarvis continued to drive for a few hundred yards while frantically contemplating on what he should or should not do, then quickly pulled over to the side of the road and reversed. The creature, however, was nowhere in sight. The encounters, however, were only destined to continue.

For example, a man named Jack Burks called me, quite out of the blue, to say that he heard stories of a big black cat roaming around the perimeter of Royal Air Force Stafford over the course of several months – way back in the late-1950s, no less. Interestingly, Burks says that a friend of his who was stationed at the military base at the time told him, 'probably around 1959 or 1960', that the RAF had opened a 'secret-file' on the sightings. This was very encouraging news indeed; unfortunately, however, repeated and in-depth attempts on my part to locate the *X-Files*-style documentation via the terms of the Government's Freedom of Information Act have not yet yielded results – positive or otherwise.

Farm-worker and fence-contractor Richard Woolley of Rugeley had his own account to relate to *The C-Files* in the summer of 2000. It was at a farm near Shooting Butts Road that Woolley would become the next witness to the activities of the Chase's resident four-legged beast. Woolley described the events as follows:

> 'My chainsaw had just run out of petrol and I was walking back to my truck when I saw it. It was the colour of a baby lion with a white tip on its tail and must have been three feet long. I was pretty scared so I got in the cab of my truck.'

When he shouted in the direction of the cat, Woolley said, it turned in his direction and merely sauntered off into the safety and cover of the undergrowth. Notably, he added that on the farm where he worked 'some of the sheep have been injured and they looked like wounds that could have been caused by a big cat. I never thought I'd see it, but I have.'

When sightings of a big cat occurred in the vicinity of the Cannock Chase Enterprise Centre in the latter part of August 2005, it prompted comments from certain concerned locals that it would surely only be a matter of time before a tragedy of overwhelming proportions occurred and a human death inevitably occurred.

One of those that had seen the animal as it prowled the woods adjacent to the Centre was warehouse worker Anthony Cooper, who recalled:

> 'It was around lunchtime. A lot of people are saying they heard something like a growl, but I didn't. I heard a rustling noise in the bushes. As soon as I turned my head I could see something in the trees – a large black shape moving slowly.'

Cooper added:

> 'It was acting as if it couldn't even see me. It just strolled past, at a distance of about ten feet, went into some thicker bushes and was just gone. I called some more people over, but as I did, they ran over telling me what they had just seen. It was an eerie feeling.'

It certainly was; and especially so for receptionist Claire Clarke, who also was in a prime position to see the big cat:

> 'It was around midday. I was walking through the woods with my boyfriend, Carl. We were talking as we walked through the woods. All of a sudden I felt him let go of my hand and he shouted, "Look!" Barely ten feet away was the unmistakable form of a large, black cat. 'We didn't know whether to run, climb a tree or stand

still. It was growling to itself as it walked along. I'm still shaking now to think of it.'

Clarke added:

'It's typical that I'd decided to leave my mobile phone on my desk at work, else I would have had a picture of it. It definitely looks like a panther. I've heard people saying that it's perhaps a domestic cat. Well, to me it looks like about seven or eight domestic cats rolled into one.'

With words that definitely stirred up the close-knit community of Cannock and its surrounding villages, she concluded: 'What if I had been an 80-year-old? Or if we had a child with us who was running around nearby trees? You can't help but think it would have been a bloodbath.' Once again, the area was gripped with cold and stark fear of what very well might be lurking in the darkness, just readying and steadying itself to pounce and feast.

March 2006 was a month that marked a true turning point in big cat activity on the Cannock Chase. A careful and in-depth perusal of local newspaper archives made that fact abundantly clear to me: 'Terrified Chase-folk are certain a panther is living in their midst after a spate of fresh sightings in the forestland around their homes,' reported the *Chase Post*. As evidence of this, the newspaper cited the testimony of Huntington-based insurance clerk Andrew Lomas, who had a close encounter of a distinctly weird kind with one of the elusive big cats late at night near the large Pye Green Tower – one of the very few telecommunication towers in Britain that is constructed out of reinforced concrete, and designed and built during the Cold War as part of the Government's radio-communications network.

Lomas stated that:

'I was riding through to Pye Green at around 11.00 p.m. when I saw something dark in the road, in my headlights. At first I thought it was a deer, but it moved so fast as I got nearer, I got the impression it was something else.'

Indeed, it was:

'As I passed the spot where it had been, I slowed down and looked into the bracken. I could make out a pair of red piercing eyes staring back at me. It gave me the chills and I drove off straight away. I never believed all these stories, but now my opinion has changed. I know what I saw.'

Echoing Andrew Lomas was the testimony of Janette Muller of Hednesford:

'I was out walking with my boyfriend over Cannock Chase when we noticed a disturbance in the bracken ahead of us. All of a sudden this big black shape jumped up and took off up the hill. I'm not sure at all what it was, but it was much too big to be a domestic animal. It made a hell of a noise as it raced away.'

The Mystery Animals of Staffordshire

In the latter part of March 2006, two truly sensational stories of encounters with big cats in the vicinity of the Cannock Chase surfaced that most definitely cannot be ignored in our search for the truth. *Chase Post* editor, Mike Lockley, received what, ultimately, may prove to be the most important one of the two – if the frustratingly-elusive evidence relating to the case ever finally surfaces as a whole, that is. The source of the account was very careful to advise Lockley that he did not want to be named – which, of course, means that the possibility of a hoax can never be fully ruled out – due to the fact that he had an extraordinary tale that he wished to impart to the newspaper.

So the story went, the man had been crossing the woods when, in the undergrowth, he noticed something truly extraordinary: nothing less than the huge skull of an unknown animal. 'This is certainly not something that you find every day,' the man told Lockley in what certainly has to be an understatement of epic proportions! He further advised the paper's editor that:

> 'It looks like it's been there for some time [and] looks like some sort of big cat to me; the fangs are enormous, and it's definitely not that of a dog or a fox.'

The man expanded yet further with a very intriguing observation:

> 'Nearby were a few other bones and a despicable trap which was secured to a tree and had a bone caught in it. I think whatever it was must have got its leg caught and then starved to death. I threw the trap away; I think they're illegal to possess.'

In the very same week that the above-incident occurred, Hednesford resident Tony Taylor stumbled across the remains of a horribly mutilated deer while walking across the Cannock Chase. Photographs taken by Taylor graphically revealed that the animal's torso had been completely picked clean of flesh, with only the head, the legs and the skeleton remaining relatively intact. Up until that point, Taylor had been completely sceptical of the theory that the Cannock Chase was home to one or more big cats. After his gruesome discovery, however, he definitely began to waiver on that position:

> 'I've been walking on the Chase for years and I've never seen a big cat,'

Taylor stated and added that:

> 'I never thought there was anything there but now I'm not so sure. Its back and ribs looked broken and the flesh was ripped from the bones. Whatever had eaten it had eaten the lot. It looked fresh, as though it hadn't been there long; I'd say no more than two days.'

A Cannock Chase Forestry Commission ranger named Jim Stewart had a very different opinion on the whole matter, however. Asserting that such discoveries were actually quite commonplace, and that the big cat accounts were completely unfounded, Stewart said:

> 'Almost certainly this deer was hit by a car and then stumbled into the woods to

die. It would have been picked clean by badgers and foxes.'

Tony Taylor was still not convinced by this particular line of theorising, however:

> 'It was about half a mile from the nearest road. I'm not sure it could have walked that far with a broken back and ribs.'

And not everyone else was convinced either; and particularly so when the reports continued to surface with unnerving and effortless regularity…

Chapter V
Staffordshire Goes Big-Cat Crazy!
'...Something ran across the road, it looked big and black...'

'The disturbing discovery of the fleshless corpse of a mutilated deer on Cannock Chase is leading many to believe that the mythical beast of Cannock Chase is back and on the prowl,'

said Mike Bradley of the *Chase Post* on April 3, 2008. Bradley added:

'Matthew Harrison, of Cannock, said that he was disturbed by the discovery. Every ounce of flesh seemed to have been stripped from the animal in a meticulous manner.'

The newspaper quoted Harrison as stating:

'Me and the family were out for a walk at Cannock Chase on Sunday March 30. As we were walking I found a carcass of a deer. It was a very fresh kill - no smell and very few flies. What flesh was left was still red and fresh. To me, if the animal had died naturally, surely it would take a while for small predators, such as foxes, to dispose of the carcass, or for it to decompose. But the most interesting thing was that when I found another carcass only 200-300m away, this one had been killed or died a few months ago. Maybe the story of wild cats on the Chase are true, and this is one of its hunting places?'

'Has beast claimed a kill on the Chase?' asked the *Chase Post's* Matt Lloyd on May 8, 2008. And Lloyd had good reason indeed to raise such a thought-provoking and sensational question. He wrote in the pages of the *Post*: 'Pye Green resident John Brazier sent us this picture from the Birches Valley area of Cannock Chase showing the mutilated carcass of a deer. And John, 54, believes it could be the work of the infamous Chase Panther.

'He came across the grim discovery on his way to work on Monday morning and says it is the second such corpse he has seen in a week. Speaking exclusively to *The Post* he said: "I saw it and thought I have to capture it, so I stopped and took the picture with my camera phone. It wasn't there on Friday and it was gone by the time I drove home from work. Something has really had a go at it. The guys at work were gob-smacked when I showed them the pictures. I've lived in the area all my live and I've never seen anything like a panther. I've seen and read a lot of articles about them on The Chase and when I saw that it makes you think. It makes you wonder if something is out there. Seeing what's happened to that deer it's got to be something of a beast. People ought to be aware if something is out there because a lot of people walk their dogs over the Chase."'

Matt Lloyd delved even further, and told the readers of the newspaper that:

'Similar discoveries in the past have been put down to cars hitting the wild animals which are then eaten by scavengers. But John is positive this animal was not hit by a car. He said: "God knows what's happened to it but it's not been hit by a car, there were no skid-marks in the road. It's happened in a really short space of time. Foxes would have chewed on the bones; the skull looked like it had been sucked clean."'

A fortnight later, specifically on May 22, 2008, Matt Lloyd wrote the following:

'Further grim discoveries have been made on Cannock Chase fuelling beliefs the beast is out there. Following the discovery of a mutilated deer near Birches Valley three weeks ago two more Chase Post readers have come forward to tell of their shocking discoveries. Both say they believe something is targeting deer in the area after finding stripped carcasses miles from roads.'

Lloyd continued with a humorous play-on-words that has been utilised by numerous, adventurous journalists hot on the trail of Britain's big cats for years: 'And the possibility the Chase Panther is a reality has given them paws for thought.' Groan...

He added:

'Nick Brown, from Tamworth, stumbled across a dead deer earlier this year during a bike ride with friends. Upon realising the find was not an isolated incident, he sent his picture to *The Post*. He said this week: "We found it quite strange at the time because the bones were perfectly clean. It didn't look weathered at all. I imagine it would have been more destroyed or pulled apart, but I don't know if you can put it down to a big cat because they probably tear and pull, too. I just don't know what it is."'

And the sightings were far from over, as Matt Lloyd amply demonstrated to his eager audience:

'Pye Green resident Julie Mason stumbled upon another carcass when out walking with her family three weeks ago at Spring Slade Lodge. She said: "We got lost and just stumbled across it. It didn't look like a hit and run because all of the joints were intact. The poor thing had been out there for some time, we are really worried as to what is happening to the animals.

My family and I have lived in the area all our lives and to see such a site is quite concerning."

Interestingly, Lloyd added:

'Julie's family [was] shocked to find the antlers of the animal apparently sawn off and its legs missing.'

This statement would seem to suggest that the killing had a far more down-to-earth explanation: namely, poachers.

But Julie was very careful to add the following:

'The bones were very clean and it just doesn't seem right how it was left like that. We looked around for the jaw and leg bones but had no joy finding anything; but I do think something is out there - I just don't know what, but to strip the deer bare it's nasty."'

Significantly, Matt Lloyd noted in closing that:

'...this may not be Julie's first encounter with the area's fabled beast: "Eight or nine years ago I was driving towards Birches Valley. Something ran across the road, it looked big and black. It looked a bit like a Labrador but the tail was big and fat. It's very strange"'

It certainly *was* very strange – and, it still is!

Careful not to be outdone by its sister-publication the *Chase Post*, on June 5, 2008, the *Stafford Post* newspaper highlighted the report of a former Royal Air Force man who had seen a big cat on two occasions in Stafford – both in 1995. The witness said that:

'the panther was a fully-grown specimen: its tail was about as fat as a deodorant-can. It wasn't at all nervous and it loped slowly across the main road as I stopped my car to watch it.'

Then, on October 30, 2008, the *Stafford Post* ran an article titled *My Encounter with the Beast* that went as follows: 'A local man has described his terrifying encounter with "The Stafford Beast." And though Philip Green only heard the creature it was enough to send him fleeing from Cannock Chase. Mr. Green, who also contacted the police last week, told the 'Post':

"The reason I'm going public with this is to make other people aware of what's out there." In recent months there have been a scattering of sightings of a "very large black cat" in Stafford Borough. And it was around 1.30pm last Wednesday afternoon that Mr. Green encountered what he believes was the beast. "I went out over the Chase and had my mum's little Yorkie with me. I parked up just past Milford and the entrance to Shugborough. We'd only gone down the track a short way when I heard the cat. I'd never given these stories about big cats any thought, but when I heard the growling there was no doubt in my mind what it was. I couldn't see this thing because there was all this bracken and ferns. But as I walked on I could hear it snarling - it was a really deep sound. There is no question in my mind what it was."'

The *Stafford Post* further reported:

'And Mr. Green did not wait to find out what the source of the growling was: "I couldn't get out of there quick enough," he said. "It sounded like I had really upset it." He added: "I don't know if this thing would attack humans - but I just want to make people aware that there is something out there."'

'Expert Plays Down Threat as Chase Panther Spotted Close to Homes' was the headline that leapt out of Staffordshire's *Chase Post* newspaper on July 23, 2009. The 'Post' informed its readers that:

'Sightings of the Chase Panther continue to rise this summer, with even more eyewitnesses claiming to have seen the beast encroaching on residential areas. Witnesses have contacted the 'Post' to say they have spotted the big cat around the Hawks Green area. The sightings follow a report last week, when residents claimed to spy a large cat-like creature roaming empty units, near housing on Hemlock Way. One woman from Heath Hayes claims she had a run-in with the beast in the same area last week: "I was walking my dog by the pool in Hawks Green," the woman, who does not wish to be named, said. "As I approached the pool, I saw, what I thought was a large black cat's tail sticking out of the bushes. At the time I thought about the articles I had read of panther sightings on the Chase. It disappeared as I approached it so I didn't see the rest of the cat. As I walked on, I kept looking to see if it was in the bushes hiding. I felt uneasy as there was no one around. I have never walked so fast as I did that day, and was glad to get home."'

The 'Post' added that one week earlier, a student named Edward Phillips reported that he had seen a panther lurking around the units of the former *Maymies* nightclub in the early hours of July 9 while he was making his way back home. Said the newspaper:

'At first, Edward said he thought he was seeing a deer, but changed his mind when he saw the way the animal moved between the units. "I have always taken these stories of sightings with a pinch of salt," Edward said. "I have heard people say they have seen some sort of a big cat around here before, but I always thought it was nonsense. But I know what I saw. It was a large black animal, and it moved very quickly. It had its belly very low to the ground, and sort of slunk its way around. I'm not saying it was a panther or anything like that, but it certainly was not anything I've encountered before."'

One location in Staffordshire that has been hit by a veritable wave of big cat encounters over the course of the last year alone, is Tamworth: a town and local government district that is located 14 miles north-east of the city of Birmingham, and which takes its name from the River Tame, which flows through the town, as does the River Anker.

An ancient locale, Tamworth has existed since Saxon times and once was the capital of Mercia, the largest of all English kingdoms of its time. The town was later sacked by the Danes - in the 9th century – and, as noted earlier, a wooden fort was constructed on the site of the current castle, designed to defend the town against further Danish invaders by Ethelfleda, Lady of the Mercians, the daughter of King Alfred the Great. In the 11th century, a Norman castle was built on the probable site of the Saxon fort, which still stands to this day, as an important and bustling tourist attraction.

Back in April 2004, the BBC reported that a number of residents of Walton, near Tamworth had claimed sightings of a large cat on the loose. Villager Richard Meredith said:

> 'There have been sixteen sightings of big cats within a mile or two of this village. [They were] sightings from people who were not drunk. They were not small pussy-cats; they were sightings from just fifty yards. Undoubtedly, there have been panthers in this area.'

It was most definitely in 2009 that matters began to heat up in the area, however. On July 30, 2009, the *Tamworth Herald* newspaper recorded the following:

> 'The mystery "big cat"' was spotted in the heart of Tamworth in the early hours of yesterday morning (Wednesday), bringing to the total number of sightings in the area to nearly 40. The creature was seen by Margaret Locke, the Musical Director of Tamworth Ladies Choir, who lives in Bonehill.'

Locke told the newspaper:

> 'I could hear this noise and I thought it was cats fighting, so I got up and went to the window and saw that it wasn't cats fighting, it was one huge cat standing on my drive, howling. It stood there for around three minutes before sauntering off. I am not given to flights of fancy and I am absolutely sure of what I saw. It was around two feet long and its tail was about the same length again, curling up at the end. I have heard cat noises on and off several times in the last week and I didn't really think anything of it. It was quite a shock to see it was a big cat. I feel sorry for it really and I would like to see it protected.'

Then, on September 28, 2009, came the following from the *Tamworth Herald*:

> 'The mystery big cat has been spotted in Tamworth again: this time on a housing estate. Lisa Urry, of Borough Road, spotted the creature on the outhouse of a nearby property in the early hours of Monday morning last week.'

In relating the details of her encounter to the 'Herald' Urry recalled:

The Mystery Animals of Staffordshire

> 'It was about 7.15am and I was looking out of my baby's bedroom window when I saw this big, black creature just standing on an outhouse about five or six houses down from me. I couldn't believe it. I just thought: "That's big for a cat." I watched it for three or four minutes and it had a wee and walked off. I think there is an aviary next to the outhouse, so perhaps it was interested in the birds. I remembered seeing a billboard about the black cat at the newsagents, so I sent my son to buy the 'Herald'. When I saw the tail of the black cat in the Herald photograph I knew that it was definitely exactly what I had seen. I never really believed in the big cat stories before, but now I'm sure.'

And the reports showed no signs of stopping at all.

> 'A villager has spoken of her shock after coming face-to-face with a big cat in a quiet country lane near Polesworth,' said the *Tamworth Herald* on October 19, 2009. The newspaper detailed the facts: 'Sue Swift, aged 45, performed an emergency stop when she saw what she thought was a large black Labrador in the road, and tried to coax the animal into her car. But minutes later, she realised that it wasn't a dog at all. It was what she believes was a puma.'

The incident, the 'Herald' stated, had occurred on a country lane between Polesworth and Dordon at around 7.30pm on Monday, October 5.

Swift told the 'Herald':

> "I saw the animal in the middle of the road. I did an emergency stop and without batting an eyelid, it walked past my car. I put my hazard-lights on and got out and opened the back door of the car. I have a very large dog myself and I just thought this was a stray dog. I started to say "come here boy", but as I have almost lost my voice, it didn't respond. I said it a few more times and then as I got closer I realised it was a lot larger than my dog. I was around eight feet away from it and I started clicking my fingers. It turned to look at me and I realised it wasn't a dog at all. It had cat eyes and I think it was a puma. When I realised, I was shaking like a leaf and very slowly I got back into my car and drove off."

The 'Herald' had even more to add:

> 'Just three days later and two miles away, a 22-year-old cyclist was stunned when the big cat bounded across the road in front of him as he cycled through Warton. Robert Piper was travelling through Warton from Polesworth at 8.30pm last Thursday when he saw the creature. The big black cat leapt across the road in front of him, its outline clearly lit by the lights from passing cars.'

Piper said:

> "I was riding my bike and I had my earphones in. A car came behind me with

> its headlights shining, so the road was lit up. The animal jumped out of the bushes just a few feet ahead of me. It bounded across the road in two jumps and disappeared into a field. It was about as big as a Labrador with a long curly tail and quite definitely was a cat. It certainly made me jump a bit. I was really shocked. I have never pedalled so fast in my life!"

And, on November 5, 2009, the 'Herald' informed its readers of a sighting that had occurred only weeks earlier, on October 13:

> 'TNT employee Gemma Capostagno was walking her partner's two Labradors on Baddesley Common when she spotted the creature. She said: "I vividly remember it, as I was talking on the phone to my partner at the time. I was just reaching the top of the 'dummy hole' when I saw a large black animal run across a section of common in front of me about five metres away. I immediately told my partner what I had seen. I only briefly saw it, as it seemed to run across the section of common I was on and then down the side of the dummy hole."'

Capostagno added:

> 'It was easily as big as my partner's dogs and I would say longer and more sleek. At first I questioned my judgement and looked around for anyone who may have been walking their dog at the time, but there was no-one to be seen. This could easily have been the mystery black cat in question.'

It could indeed…

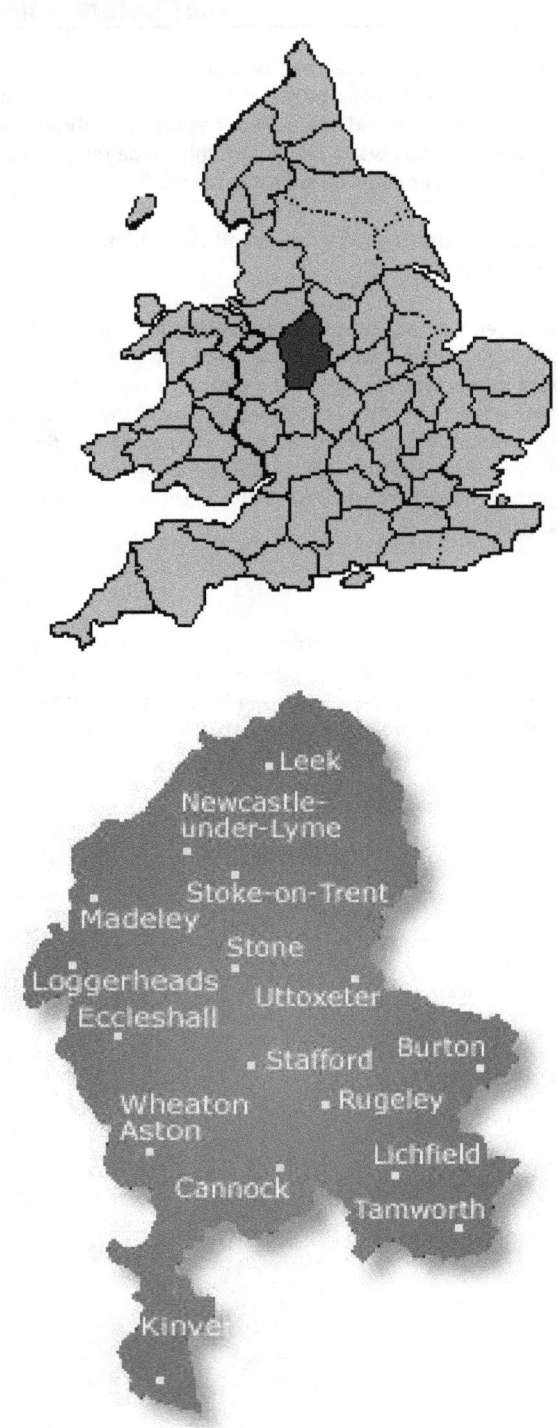

Chapter VI
From Where Do The Cats Come?
'...Breathe deep, pull hard...!'

If there were, and almost certainly still are, big cats living and thriving in stealth throughout much of the county of Staffordshire - as the evidence does strongly suggest is so - then the biggest question of all surely has to be: from where on earth did they originate? It is a little-known fact that encounters with large, predatory cats in Staffordshire actually long pre-date the current obsessions with the Beast of Bodmin, the Beast of Exmoor, and their ever-increasing exotic ilk.

For example, only nine miles from the fringes of the Cannock Chase and specifically near the town of Brewood, stands Chillington Hall. The present hall is actually the third one: a castle was built on the site in the 12th century, while today's hall was constructed in 1724. It is, however, the *second* Chillington Hall that specifically concerns us. It was there, in the 1500s, that one of the first private zoos was established – by nobleman Sir John Giffard. According to local legend, on one fateful day, Sir John's favourite animal, a leopard no less, escaped from the confines of its enclosure and charged headlong into the wilds of the surrounding Staffordshire countryside. Arming himself with a cross-bow, Sir John, along with his son, quickly set off in hot pursuit of the marauding animal; and, to his total horror, found it poised to attack a terror-stricken mother and child who were cowering on the ground.

In an instant, according to the old story at least, Sir John drew his bow and took careful and quick aim. At that very moment, his son cried out: 'Prenez haliene, tirez fort!' or: 'Breathe deep, pull hard!' Sir John sensibly, and rapidly, took his son's advice and fired. With but just one shot, the leopard fell to the floor, utterly stone dead. Giffard's Cross – which still stands to this day – was raised where the creature is reputed to have taken its last breath; meanwhile Sir John decided it might be a very good idea to adopt his son's words as the family's motto.

Of course, if one large, exotic cat was roaming the wilds of Staffordshire as far back as the 1500s, then who knows how many other possible escapees there might well have been that *weren't* cut down by the power of Sir John's mighty bow-and-arrow? It is a sobering thought, indeed, to think that there may possibly have been large, wild cats living stealthily in the woods and forests of Staffordshire more than five hundred years ago – and perhaps even reproducing and thriving, too. Perhaps the mystery of Staffordshire's big cats is a far, far older one than we generally give it credit for. And there was far more to come, too.

More than a century ago, a zoologist – Dr. John Kerr Butter - kept a veritable menagerie of wild and exotic animals in his back-garden, which stood on the grounds of what is today Cannock's Wolverhampton Road Police Station. Indeed, Butter's neighbours were regularly confronted by a truly wild assortment of animals craning their necks over his garden fence – including, astonishingly enough, nothing less than a fully-grown giraffe!

Interestingly, records reveal, Butter had personally raised and tamed a wild cat – specifically an ocelot - a feat then considered impossible by the zoological elite; and something that earned the doctor the distinction of being made a Fellow of the Royal Zoological Society. And, somewhat amusingly, patients at Dr. Butter's office would often find themselves sat in the waiting room next to his favourite monkey, Antony, who had practically turned Butter's practice into a definitive home-from-home.

Not only that: Butter had also amassed a large collection of artefacts in his laboratory, including everything from bear-skins to alligator-jaws that had been made into pen-and-ink stands, as well as jars of carefully-preserved animal-organs. Butter was also held in very high regard by the populace of Cannock. For example, when the Boer War broke out, the doctor immediately elected to do his duty for his country, and on the day of his departure for the war-torn battlefield, recorded the *Chase Post*, 'nearly the whole town gathered to see him off'.

Said local historian, Dave Battersby: 'He was the recognised doctor in the town and was very well known many years after his death. He obviously did a lot of good for a lot of people. He trained people in first-aid.' The doctor did indeed; and he also established the very first ambulance service across the Cannock Chase. But there was a far more significant development still to come. With the outbreak of the First World War in 1914, food supplies in the Cannock area dwindled dramatically and drastically and Dr. Butter was unfortunately forced to let his animals go.

In the period from 2007 to 2009, I spoke with several Staffordshire-based historians and consulted the available newspaper reports of the day, but I did not uncover any real indication as to what had specifically happened to the doctor's wide and varied animal collection. Since, however, there have been no reports of giraffes running wild in the woods of the Cannock Chase, it seems very safe to assume that this particular beast was successfully re-housed somewhere else. But about Antony, the little monkey, and the tamed ocelot, no-one knew anything at all.

Historian Dave Battersby added that: 'It's likely they were given to other collectors or to zoos. The moving of animals wouldn't be newsworthy at the time and so archives won't tell us much about where the animals went.'

The Mystery Animals of Staffordshire

Indeed, there still is no available evidence to suggest that the creatures had been donated to new zoos. But, equally, there is no evidence to suggest that the animals were put to sleep, either. They had just vanished – completely and utterly, it must be said. That is, unless the good doctor had, perhaps, decided that the wisest approach to solving the problem was to clandestinely release them into the wilds of Staffordshire late one night when everyone else was asleep and tucked up in their beds.

If he *had* done so, and if there were several such animals that subsequently successfully bred, then that would very possibly go some significant way towards possibly explaining the presence of large, wild cats on the Cannock Chase, and throughout the rest of Staffordshire, today. And encounters with Antony's offspring, and their offspring, might also assist in explaining a few of the truly baffling sightings of monkey-like animals throughout Staffordshire – as will become far more than apparent in a subsequent chapter.

But there are still yet other explanations for the presence of the many and varied out-of-place big cats throughout Staffordshire.

If someone had said to me before I embarked upon my quest for the truth about the big cats of the region that I would find myself digging into the accounts of a man known as the Lion Man and his pal – whose moniker made him sound like something straight out of the pages of *Treasure Island* - One-Eyed Nick, I would probably have merely smiled and forgotten all about it. But sometimes truth really *is* stranger than fiction. Dudley's *Lion Man* – or Louis Foley, to give him his real name – claimed to have been personally acquainted with a number of people who had stealthily released big cats into the heart of the Cannock Chase in the 1970s; largely as a result of the significant changes that were made to the previously-mentioned Dangerous Wild Animals Act.

Foley's fascination with big cats began more than three decades ago – when he purchased his first lion, ostensibly as a 'guard dog'. Astonishingly, his interest eventually resulted in him possessing an absolute zoo of exotic big cats, including no less than seven lions, plus tigers, pumas, panthers - and nothing less than a crocodile, too!

According to Foley, 'Heartless cowards who bought panthers and other big cats as fashion accessories soon realised what a handful they could be. They were left to die in areas like the Chase and many of them would have perished because they were tame. But I have seen tracks and evidence of kills that proves there are many that survived.'

Foley added, somewhat guardedly, that his friend, 'One-Eyed' Nick Maiden, personally released both a panther and a cougar, after they had been given to him when their owner became completely unable to cope with caring for the beasts.

Foley said: 'I was away at the time and I was furious, but 'One-Eyed' Nick had no choice; we just did not have any more room. I would never release a "tame" big cat because I have respect for them and keeping one is like having a child. You have a responsibility to look after them. When I heard about them being released, I would travel in the area and try to recapture them in case they could not survive.'

In conclusion, Foley stated:

> 'If people could no longer look after the big cat or could not afford to keep them – and did not want them put down – an area like the Chase would be an ideal place to release one. I lived with my big cats for years and I would assure Chase folk there is no danger to their lives. Wild cats will avoid humans.'

In the same week of March 2006 that the details of Tony Taylor's big cat encounter, as described in an earlier chapter, were made public, a truly sensational story surfaced – again from the town of Hednesford – from a woman described by the *Chase Post* newspaper as 'Miss S. Thomas', who claimed that she had raised a big cat from a cub, no less than eleven years earlier.

Stating that she had previously been reluctant to come forward for fear of ridicule, Thomas informed the *Post* that she had been roaming on the Chase back in 1995 when:

> '...I saw this little black cat lying in the undergrowth. It looked completely lost, as if it had been dumped and we took pity on it and took it home.'

Initially at least, there seemed to be nothing untoward about the situation, at all. As time progressed, however, matters took a distinctly strange turn, and that is putting matters extremely mildly. The cat began to grow at a highly alarming rate, said Thomas, and developed some extremely disturbing character traits:

> 'It really started to put on weight and quickly was larger than a normal cat. It also started to act weird. It wouldn't lie down like a domestic cat and it didn't purr normally; it sort of growled. My friends all told me that it was not a house cat and I had my suspicions.'

Matters took a *very* severe turn for the worse, to say the least, when the cat turned its animalistic attentions towards Thomas' daughter:

> 'The thing attacked my daughter when she was asleep and left her with a nasty cut. That's when I decided to throw it out. I took it to live outside, but it soon disappeared. I presume it went back to the Chase.'

She concluded:

> 'I often wondered if that was the big cat that everybody is talking about. I thought if I told the story people would think I was mad, so I kept quiet.'

We will probably never know with absolute certainty if the cub that Thomas found *had* been quietly dumped on the Cannock Chase by someone who was not in a position to care for it; or even if the story was itself nothing more than a very well-constructed, ingenious and audacious hoax. But if not, then this suggests yet another possibility. And it is a highly disturbing possibility, too: namely, that Britain's big cats have not just made a home for themselves in Staffordshire; they are breeding throughout the county, too…

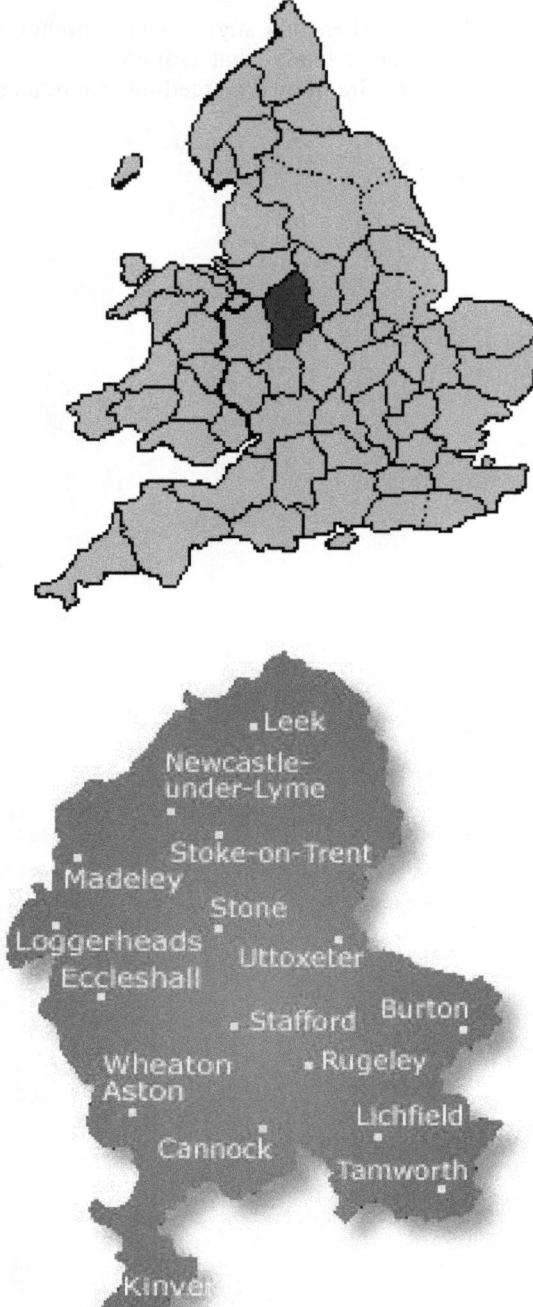

Chapter VII
The Taigheirm Terror

'...After a certain continuance of the sacrifice, infernal spirits appeared in the shape of black cats...'

I am personally convinced that the overwhelming majority of all of Staffordshire's big cats – and those that are said to roam the rest of the British Isles, too - are wholly flesh-and-blood animals that have successfully managed to survive and thrive in the county's wide and varied countryside for decades; and perhaps for even longer too, if some of the Staffordshire-based cases that have been cited already are anything to go by. There are, however, a few incidents that, for some people who have been exposed to the controversy, seem to suggest far stranger points of origin for at least *some* of the cats in question – albeit, certainly not all of them.

As I have noted earlier, there have been several sightings of big cats (not to mention other animal-based anomalies, too) in the vicinity of the Cannock Chase's German Cemetery. And, on this matter, I am aware of at least three cases from this particular location where the sightings seem to have involved anything *but* flesh-and-blood big-cats.

For example, Eileen Allen says that she caught sight of a 'big black panther', as she described it, while she was visiting the cemetery in the latter part of 1996. The overwhelming shock of seeing the immense beast lurking near one of the head-stones, and staring intently in Allen's direction, was nothing compared to the absolute terror that struck her when the creature suddenly vanished – and I do mean literally vanished: into thin-air - amid a sound that Allen described as 'like an electricity cracking noise'. Unsurprisingly, Allen did not hang around and quickly left the cemetery. To this very day, she has never returned; nor does she have any plans to do so in the future, either.

Bob Parker experienced something very similar in the woods barely a quarter of a mile or so from the cemetery in late 2000, while walking his dog on one particular Sunday morning. In this case, a large black cat came hurtling violently through the heather, skidded onto the pathway that Parker was following, and bounded off, apparently not at all bothered or concerned by the presence of either Parker or his little Corgi dog, Paddy. Of course, seeing a big cat was astounding enough in itself; but what happened next was just downright bizarre.

Parker says that: 'Me and the dog just froze solid. I couldn't believe it; could *not* believe it. But when [the cat] got about fifty feet from us, it literally sort of dived at the ground. It sort of took a leap up and almost dive-bombed the path, and went right through and vanished, just like that. I know exactly how it sounds, so don't tell me, because I had a belly-full from the missus about it all for ages. But that's exactly what happened: it was like it just melted into the path.'

Equally as strange, but for very different reasons, is the story of Sally Ward. She claims that back in the late-1960s, while she and her husband were walking across the Chase, not too far from the green and pleasant Milford Common, on what was a wintry and very foggy morning, they almost literally stumbled upon what she described as 'a black panther; a real one' that was sitting 'bolt-upright' on an open stretch of ground to their left, and approximately thirty feet from them. But that was not the strangest spectacle, however. Also stood upright around the huge beast was what Ward described as: '...seven or eight other cats. But normal cats: pet cats.'

She was sure that the smaller animals were not 'big-cat kittens', but were 'the sort of cats anyone would own. I don't know anything about panthers and lions and tigers, but I know a normal cat when I see one, thank-you very much'. Rather strangely, all of the smaller cats were staring, in almost hypnotic fashion and in complete silence, at the large black cat, as if utterly transfixed by its proud stance and powerful, muscular presence.

The Wards, perhaps quite naturally, felt very ill at ease with the whole surreal and eerie situation, and both slowly and cautiously continued past the group, and then raced down to their car, which they had parked in Milford. Ward's husband was, at first at least, fully intent on reporting the encounter to the local constabulary; but after Mrs. Ward pleaded otherwise, they decided to remain silent – aside from quietly confiding in various friends over the years and decades. To this day, Sally Ward stands by her story with total, firm conviction: 'No-one will ever tell me there isn't something funny about black-panthers in England.'

Of course, maverick (and, admittedly, very rare) cases such as the three cited above are, for many big cat researchers in Britain, the absolute bane of their subject – simply because they suggest at least some of the nations' mystery cats *may* be much more, or paradoxically much *less*, than mere flesh-and-blood animals. And, for that reason, investigating them can be highly problematic in the extreme. It is one thing to say that big cats are prowling around the wilds of Staffordshire; but it is quite another thing altogether to suggest that some of the creatures may very well have paranormal or supernatural qualities attached to them, too.

So, therefore, how do we even begin to explain and reconcile such reports in Staffordshire? Indeed, *can* we even explain and reconcile these accounts to any meaningful degree at all? The answer to that question is: well, maybe. But to do so, we have to first go back in time to 1976 and head off to the counties of Devon and Cornwall. And then, having done that, we will take a wild ride even further back in time: specifically to Scotland's Isle of Mull and a series of infernal rites and rituals that may possibly, and ultimately, go some way towards explaining the small number of truly strange and anomalous big cat encounters in Staffordshire – and maybe even elsewhere, too.

As practically each and every British soul aged about forty and upwards will graphically recall, midway through 1976 the good folk of the nation were pummeled into exhaustion and submission by what was one of the hottest summers on record for many a decade. Water-supplies came perilously close to completely drying-up; people were seemingly dropping like flies from the effects of overwhelming heatstroke; and there were even rumours flying around that the Army would take control of the streets in the event that the fabric of society should begin to unravel and crumble.

Fortunately, such an apocalyptic scenario did not come to pass, and the extraordinary weather eventually gave way to what generally passes for normality in the British Isles: namely, rain, wind and dark grey skies. And everyone, I am sure, breathed a distinct sigh of relief when the sun finally elected to move on to pastures new. That same year was famous, or rather infamous, for a whole variety of things beyond just the bizarre weather, however.

It was also in 1976 that one of the most monstrous of all cryptozoological creatures loomed wildly out of the woods that surround Mawnan Old Church in Cornwall: the Owlman. A diabolical, glowing-eyed and winged entity that could very easily pass for Mothman's twin-brother, the beast became the absolute talk of Cornwall for months, and ultimately sent the Centre for Fortean Zoology's Director (and the editor and publisher of this very book, to boot) Jonathan Downes on a dark and disturbing quest that came perilously close to utterly destroying him, both mentally and physically.

And if that was not nearly enough, from the depths of the waters off the coast of Cornwall surfaced a veritable serpent of the deep: Morgawr, the Sea-Monster; truly a southwest equivalent of the Loch Ness Monster, if ever there was one. But there was far more high-strangeness afoot in this same, precise time-frame and locale, and absolutely none of it was of a positive nature at all. And none was able to better chronicle this strange affair than Jon Downes himself:

> 'Cats were disappearing from Redruth, Falmouth, and Penryn over the summer. According to a representative from the Cat's Protection League, abnormally large numbers of cats were missing from their homes. The spokesman for the charity claimed that they were being sold to vivisectionists; but that is a common story used to explain such spates of missing creatures, and is I feel, unlikely to be true.'

Downes continues:

> 'Cats are among the easiest creatures in the world to breed. There is never any shortage of kittens free to a good home, and were vivisectionists, or cat-furriers, to want a supply of such animals, it would be extremely easy to arrange a constant supply without the need of breaking the law. It is also unlikely that the market forces would demand a particularly large number of such animals; and it is almost certain that any hospital, research laboratory, or chemical firm worth

its salt would buy their experimental animals from reputable suppliers rather than from the sort of shady character who would attempt to make a living from selling stolen pears.'

As Downes also astutely notes:

'Spates of missing cats almost certainly have a different explanation; but they are of great interest to the Fortean. It is a well known fact that the number of missing cats rises according to the number of sightings of alien big cats [known as ABCs] in the vicinity. Zoologists who believe in such things will claim that this is because pumas – the animals most likely to be responsible for the vast majority of "genuine" (whatever that means) ABC sightings, are partial to the taste of their smaller, domesticated cousins.'

There is, however, another explanation for the sightings of giant, mystery black cats and their associations with, and connections to, everyday house-cats. It is an explanation that takes some very strange twists and turns of a definitively disturbing nature, and has at its black heart a vile and disturbing conspiracy that extends across the centuries, throughout the higher echelons of power, and that is truly evil in nature. An ancient rite borne out of darkest and ancient Scotland, it is called the Taigheirm.

British big-cat researcher Merrily Harpur says of the Taigheirm:

'This was an infernal magical sacrifice of cats in rites dedicated to the subterranean gods of pagan times, from whom particular gifts and benefits were solicited. They were called in the Highlands and the Western Isles of Scotland, the Black-Cat Spirits.'

The rites in question, Harpur added,

'...involved roasting cats to death on a spit, continuously for four days and nights, during which period the operator was forbidden to sleep or take nourishment, and after a time infernal spirits would appear in the shape of large, black cats.'

Without doubt, the most detailed description of this archaic and highly disturbing rite can be found within the pages of J.Y.W. Lloyd's 1881 book, the truly monumentally-titled: *The History of the Prince, the Lords Marcher, and the Ancient Nobility of Powys Fadog and the Ancient Lords of Arwystli, Cedewen, and Meirionydd*. Since Lloyd's words are absolutely vital to understanding and appreciating the weird story that will soon unfold in its unsettling entirety, I make no apology for quoting the author in full:

'Horst, in his *Deuteroscopy*, tells us that the Highlanders of Scotland were in the habit of sacrificing black cats at the incantation ceremony of the Taigheirm, and these were dedicated to the subterranean gods; or, later to the demons of Christianity. The midnight hour, between Friday and Saturday, was the authentic time for these horrible practices and invocations; and the sacrifice was continued four whole days and nights, without the operator taking any nourishment.'

On this matter, Horst himself said:

'After the cats were dedicated to all the devils, and put into a magico-sympathetic condition, by the shameful things done to them, and the agony occasioned to them, one of them was at once put alive upon the spit, and amid terrific howlings, roasted before a slow fire. The moment that the howls of one tortured cat ceased in death, another was put upon the spit, for a minute of interval must not take place if they would control hell; and this continued for the four entire days and nights. If the exorcist could hold it out still longer, and even till his physical powers were absolutely exhausted, he must do so.

'After a certain continuance of the sacrifice, infernal spirits appeared in the shape of black cats. There came continually more and more of these cats; and their howlings, mingled with those roasting on the spit, were terrific. Finally appeared a cat of a monstrous size, with dreadful menaces. When the Taigheirm was complete, the sacrificer demanded of the spirits the reward of his offering, which consisted of various things; as riches, children, food, and clothing. The gift of second sight, which they had not had before, was, however, the usual recompense; and they retained it to the day of their death.'

Lloyd took up the story from there:

'One of the last Taigheirm, according to Horst, was held in the island of Mull. The inhabitants still show the place where Allan Maclean, at that time the incantor, and sacrificial priest, stood with his assistant, Lachlain Maclean, both men of a determined and unbending character, of a powerful build of body, and both unmarried. We may here mention that the offering of cats is remarkable, for it was also practiced by the ancient Egyptians. Not only in Scotland, but throughout all Europe, cats were sacrificed to the subterranean gods, as a peculiarly effective means of coming into communication with the powers of darkness.

'Allan Maclean continued his sacrifice to the fourth day, when he was exhausted both in body and mind, and sunk in a swoon; but, from this day he received the second-sight to the time of his death, as also did his assistant. In the people, the belief was unshaken, that the second-sight was the natural consequence of celebrating the Taigheirm.'

At this point in his narrative, Lloyd elected to quote Horst:

'The infernal spirits appeared, some in the early progress of the sacrifices, in the shape of black cats. The first who appeared during the sacrifice, after they had cast a furious glance at the sacrificer, said – Lachlain Oer, that is "Injurer of Cats." Allan, the chief operator, warned Lachlain, whatever he might see or hear, not to waiver, but to keep the spit incessantly turning. At length, the cat of monstrous size appeared; and, after it had set up a horrible howl, said to Lachlain Oer, that if he did not cease before their largest brother came, he would never see the face of God. Lachlain answered, that he would not cease till he had finished his work, if all the devils in hell came. At the end of the fourth day, there sat on the end of the beam, in the roof of the barn, a black cat with fire-flaming eyes, and there was heard a terrific howl, quite across the straits of Mull, into Morven.'

Lloyd then continued again himself:

'Allan was wholly exhausted on the fourth day, from the horrible apparitions, and could only utter the word "Prosperity." But Lachlain, though the younger, was stronger of spirit, and perfectly self-possessed. He demanded posterity and wealth, and each of them received that which he has asked for. When Allan lay on his death-bed, and his Christian friends pressed round him, and bade him beware of the stratagems of the devil, he replied with great courage, that if Lachlain Oer, who was already dead, and he, had been able a little longer to have carried their weapons, they would have driven Satan himself from his throne, and, at all events, would have caught the best birds in his kingdom.

'When the funeral of Allen reached the churchyard, the persons endowed with second-sight saw at some distance Lachlain Oer, standing fully armed at the head of a host of black cats, and every one could perceive the smell of brimstone which streamed from those cats. Allan's effigy, in complete armour, is carved on his tomb, and his name is yet linked with the memory of the Taigheirm.

'Shortly before that time also, Cameron of Lochiel performed a Taigheirm, and received from the infernal spirits a small silver shoe, which was to be put on the left foot of each new-born son of his family, and from which he would receive courage and fortitude in the presence of his enemies; a custom which continued till 1746, when his house was consumed by fire. This shoe fitted all the boys of his family but one, who fled before the enemy at Sheriff Muir, he having inherited a larger foot from his mother, who was of another clan. The word Taigheirm means an armoury, as well as the cry of cats, according as it is pronounced.'

In 1922, Carl Van Vechten commented on this particularly nightmarish ritual in a footnote contained in his book *The Tiger in the House*. It read:

'The night of the day I first learned of the Taigheirm I dined with some friends who were also entertaining Seumas, Chief of Clann Fhearghuis of Stra-chur. He informed me that to the best of his knowledge the Taigheirm is still celebrated in the Highlands of Scotland.'

Keeping this latter point firmly in mind, consider the following question from Merrily Harpur:

'Could the "Black Cat Spirits" which the Taigheirm produced be linked to the presence of huge black cats seen on Mull in the present day?'

Certainly, for such a relatively small and remote place - and one that is, of course, wholly separated from the mainland - Mull has experienced a number of significant and memorable sightings of big black cats over the course of the last few decades. Investigative researcher and writer Glen Vaudrey states in his book *The Mystery Animals of the British Isles: The Western Isles*, that one such sighting was made in 1978 by a Mrs. G.W. Brodie and a friend.

Vaudrey relates the details:

> 'When she described the sighting in 1985, she would state that at first glance they thought the animal was a black dog; perhaps a Labrador. But then it crossed the road, and started to climb over some rocks heading up to the foothills of Ben More. As it moved, it became apparent that the animal was not a dog after all, but a large black cat.'

Two decades later, adds Vaudrey:

> '...a couple [was] driving around the north side of Loch na Keal where they had been filming otters. As they drove along, they became aware of a large cat sitting upon a ridge above the road. They first noticed it as it looked down in their direction. They stopped the car and attempted to stalk the cat, but to no avail – it was having none of it, and quickly moved off into the trees.'

Notably, the very same couple had a *second* sighting only a couple of months later: this time the location was the south side of Loch na Keal, and as the beast was attempting to stalk a group of sheep – to no real or meaningful avail, as it seemingly transpires.

Vaudrey correctly states:

> 'Two reports from the area seem pretty extraordinary, but how about a third? This one was made by a group of three people holidaying on Mull in October 2003. The three of them were driving through Gruline heading back from the Ross of Mull at around 3.00 p.m. For those who are not familiar with the geography of the Isle of Mull, Gruline is to be found at the head of Loch na Keal. Do you see a pattern forming here?'

As far as the nature of the beast itself was concerned, Vaudrey explains:

> 'What they reported seeing was an animal that they described as looking like a large cat, which was either very dark or black in colour, that was stalking along the grass verge at the side of the road...The consensus of opinion was that it appeared to be a puma...'

And with respect to the Taigheirm and the big cats of Mull, Vaudrey says:

> 'Perhaps it is an echo of that act that has created the mystery big cat of Mull.'

Of course, this is a fascinating story; but what on Earth does it have to do with sightings of apparently paranormal big cats in Staffordshire? Now, it is time for us to get to the heart of the story.

In the latter part of May 2009, I received a late-night, transatlantic telephone call from a man named Donald Johnson, who related the barest of fragments of a deeply controversial story which suggested that the disciples of the Taigheirm were still up to their old tricks – but in the heart of Staffordshire, no less, rather than on the Isle of Mull; and that, he, Johnson knew certain, integral parts of the fantastic story. And so it was that on a trip back to England only three months after Johnson's call (I was due to speak at the *Weird Weekend* conference that the Centre for Fortean Zoology holds in Woolfardisworthy, Devonshire every year, and also at the Warminster, Wiltshire-based *Weird 09* gig), I was able to dig even further into the black heart of the mystery itself.

The meeting with Johnson occurred three days before my journey down to Devonshire, in a pleasant old tavern in the Staffordshire town of Rugeley. Rather significantly, and interestingly, for someone who had such an extraordinary tale to tell, there was absolutely nothing out of the ordinary about Johnson, at all: he worked as a sales-representative for a furniture company, was married with three children, and spent much of his spare time following two particular hobbies: collecting Gerry Anderson-based memorabilia and cultivating cacti. And that was about it, really. It didn't, however, take away the undeniable fact that Johnson had a tale of monumental proportions to divulge.

Basically, the story went like this: for a considerable number of years, Johnson's now-deceased father had worked for a particularly esteemed and wealthy Staffordshire family; and, over time, had worked himself up to a position of some standing in its ranks. It was around 1982, however, that something strange, memorable, and undeniably disturbing, occurred.

Johnson's father was invited to attend a Saturday night cocktail-party held at a large, country-house-style abode that was reportedly situated just outside of Stoke-on-Trent, and where a highly unusual, and very near-Faustian, offer was carefully put to him. Several of his father's close colleagues were also in attendance, Johnson explained, and whose manner suggested that they were already fully aware of the nature of the offer that was about to be placed upon the old, oak table. And thus it was that after the sun set and the group of about twelve or thirteen retired to a large and sprawling lounge, the reason for the invite became graphically clear.

Rather casually, Johnson's father was asked words to the effect of: 'Have you ever heard of the Taigheirm?' He replied that, no, he had not – which is not at all surprising, of course. As a result, there followed (a) a lengthy and detailed history of the Taigheirm; (b) an explanation of how the cat sacrifices were still being undertaken – at secure, well-guarded locations in Staffordshire, no less; and (c) dark revelations concerning their purpose, and how, allegedly, if the sacrifices were conducted strictly according to the ways of the old and now-largely-forgotten gods of ancient Britain, the resultant effect could be considerable wealth, and a high degree of both power and influence for the successful conjurer.

To his credit – or not, depending upon your own, personal perspective of the offer – Johnson's father was utterly appalled at the very idea of it all and immediately said that he wanted no part of what, quite justifiably too, sounded very much like 'devil worshipping in one of them old *Hammer* films'.

In response, the group merely thanked him for listening to their offer, assured him ('only half-jokingly', according to a deadly-serious Donald Johnson) that he need not worry about 'anything happening to him, like a car accident', and relative normality was restored via a few more drinks and discussions of a distinctly different, and far more down-to-earth and much lighter, nature.

Despite the fact that Johnson's father worked for another four years with several of those same people who were in attendance at the party, the black and disturbing matter of the Taigheirm was never again brought up with him – much to his relief, says Donald Johnson. And it was not even until the early 1990s that Johnson, himself, was guardedly told the story by his then-terminally-ill father, who finally decided to spill the beans.

But, as I had suspected would be the case, Johnson had come to one – perhaps inevitable – conclusion: namely, that at least *some* of the curiously-elusive big black cats said to be roaming Staffordshire were not flesh-and-blood entities at all. Rather, Johnson suspected strongly, their collective beastly presence was due to the resultant effects of Taigheirm-like ceremonies that were still being secretly conducted by powerful men and women all across Staffordshire and under the safety blankets of night-time and darkness, and for reasons directly relevant to personal, financial and sexual gain.

Johnson claimed intimate knowledge of several people allegedly engaged in the present day rites and sacrifices in Staffordshire – one of who, he said, was a well-known figure in 'the county's politics' in the late 1980s – but declined to identify any of them by name; somewhat conveniently, some might say. Johnson does state, however, that the truth of the matter – as well as complete and utter vindication for his story - is destined to surface no later than the allegedly-apocalyptic year of 2012: the year by which, he estimates, he will have completed a full-length book on the matter – to be titled *Supernatural Sacrifice*, and that will finally reveal the truth about the big cats of Staffordshire and their perceived paranormal origins, as well as masses of additional data on an incredible network of active Taigheirm disciples that are said to be secreted at additional locations all across the British Isles, or so Johnson earnestly maintains. Time, and Johnson's *Supernatural Sacrifice*, it seems, will tell…

Glen Vaudrey
My turn again.

It's good to see one of my rather-worth-tracking-down books quoted in this chapter but there is one small detail missing from the report that in many ways devalues the Taigheirm tale and that is the name of that most fearsome demon cat called up by this unhealthy magic, the name? Big Ears. Whether or not his side kick Noddy is in tow is not clear but really Big Ears sounds a name more suited to a cuddly rabbit. You will find out later in this volume that not all rabbits in Staffordshire are that friendly.

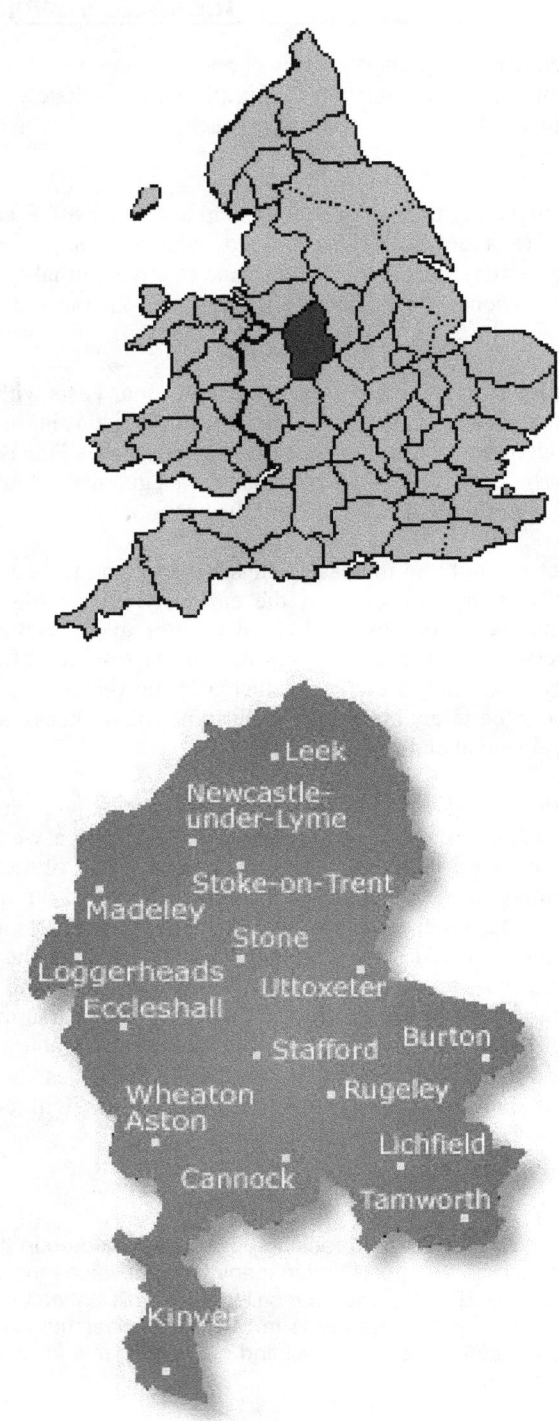

Chapter VIII
Creatures of the Castle Ring
'...It is said that the ancient horsemen of old are now seeking revenge...'

While reports of weird creatures literally abound throughout the length and breadth of the county of Staffordshire, there is one specific area of the region that certainly seems to act as an absolute magnet for such high-strangeness: it is called the Castle Ring. Located in the village of Cannock Wood and inhabited around A.D. 50 by the Celtic Cornovii tribe, the Castle Ring is an Iron Age structure commonly known as a Hill Fort. It is 801 feet above sea level; its main ditch and bank enclosure is fourteen feet high; and, at its widest point it is 853 feet across. Very little indeed is known about the mysterious and long-forgotten tribe of people who constructed the Castle Ring all those centuries ago, except to say that its creators were already in residence at the time of the Roman invasion of England and remained there until approximately A.D. 50.

Beyond any shadow of a doubt whatsoever, the absolute strangest report on record of a weird creature seen at the Castle Ring is that of a woman named Pauline Charlesworth. According to Pauline, it was a bright, summery day in July 1986 that her strange encounter occurred. As she worked on Saturdays, Pauline explained to me when we met in 2001, she had a regular day off work during the week, and had chosen this particular day to prepare a picnic-basket, and take a trip up to the Castle Ring.

On arriving, she prepared herself a place to sit, stretched out a blanket on the ground and opened up her picnic-basket that contained drinks, fruit and sandwiches. For more than an hour she sat and read, but then something very curious happened. It was as if, Pauline explained, she was sitting within the confines of a vacuum and all of the surrounding noises, such as the birds whistling and the branches of the trees gently swaying, stopped - completely. Pauline also said that 'what was there wasn't quite right'. By that statement, she explained: 'The best way I can describe it is to say it was like I wasn't really on the Chase but it was as if I was in someone's dream of what the Chase should look like: as if it was all a mirage; but a good one.'

Then, out of the trees, came a horrific form running directly towards her. It was, said Pauline, a man. The man, however, was quite unlike any that she had ever seen before. He had long, filthy hair, a matted beard, and a 'dumpy' face that was far more prehistoric than modern in appearance. He was relatively short in height, perhaps no more than five feet two inches, and was clad in animal skins that extended from his waist to his knees, and with a long piece of animal skin that was draped over his right shoulder. In his right hand, the man held what were undoubtedly the large antlers of a deer that had been expertly fashioned into a dagger-like weapon that looked like it could inflict some very serious damage indeed.

Pauline said that it was very difficult to ascertain who was more scared: her or the man. While she stared at him in stark terror, he eyed her curiously and in what Pauline described as a disturbing and sinister fashion. On several occasions he uttered what sounded like the words of an unknown language: 'It was like he was angry and firing questions at me,' she added. But that was not all.

In the distance, Pauline could hear other voices getting ever closer and closer and that collectively and ultimately grew into a literal crescendo. And then she found out the source of the noise: through a break in the trees came perhaps thirty of forty more similarly clad people, some men and others women and all chanting in an unknown and presumably ancient tongue.

It was soon made clear to Pauline that some sort of significant ceremony was about to take place inside the Castle Ring - and she, no less, was right in the heart of the brewing action. The men and women all proceeded to sit down at the edges of the Ring. One man, much taller than the rest and who she assumed was the 'leader of the group', marched over to her and said something wholly unintelligible; but that she understood by the curt wave of his arm meant that she should get out of the circle. This she quickly did and retreated with shaking legs to the tree-line. For more than fifteen minutes she sat, transfixed with overwhelming terror by the sight, as this curious band of people continued to chant and sway in rhythmic, hypnotic fashion.

Then, out of the sky, came the most horrific thing that Pauline had ever seen in her entire life. It was, she recalled, a creature about four feet in height, human in shape, with oily, greasy black skin, thin arms and legs and a pair of large, bat-like leathery wings. And, just for good measure, it had two hideous, red, glowing eyes, too. 'It was like the devil,' recalled Pauline, perhaps with a high-degree of understandable justification.

The creature slowly dropped to the ground and prowled around the Ring for a minute, staring at one and all and emitting hideous, ear-splitting shrieks. Suddenly, seven or eight of the men pounced on the creature, wrestled it to the ground, and bound it firmly with powerful ropes. It writhed and fought to get loose and tore into the flesh of the men with its claws; but was finally subdued and dragged into the forest by the same tribe-members. The remainder of the party followed and Pauline said that the strange atmosphere began to lift and the area eventually returned to its original normality. For several minutes she stood her ground, too afraid to move, but then finally returned on still-unsteady legs to her blanket and quickly scooped up both it and her picnic basket and ran to her car. Of course, the sceptic would say that Pauline's unearthly experience was merely the result of a bizarre dream – or the absolute worst of all nightmares, perhaps. And maybe that really is all it was. Nearly a quarter of a century later, however, Pauline herself is still convinced that something very strange and diabolically evil occurred on that summer's day in long-gone July 1986, and that she was provided with a unique glimpse into Staffordshire's ancient past. She may well just be right.

In October 1995, the *Cannock Mercury* newspaper reported on a series of very weird events that were then occurring in the midst of Beaudesert Old Park, which is situated very near to the Castle Ring. Home to a camp for both scouts and guides, Beaudesert was the site of repeated, strange activity in the late summer and early autumn of 1995. Not only that: a group of scouts was going to stake-out the area in a careful and dedicated bid to hopefully try and get to the bottom of the mystery, once and for all.

The 'Mercury' noted:

> 'Wardens and assistants have reported strange noises, screams and eerie goings-on around the camp.' Indeed, they had: including encounters with a dark-cloaked figure and a ghostly child roaming the woods and haunting the roads.
>
> Steve Fricker, the assistant-leader of the 2nd Rugeley Hillsprings Scouts related at the time: 'It is said that the ancient horsemen of old are now seeking revenge for the disturbances they have had to face for several years from these excitable youths.'
>
> The newspaper added: 'Scouts will be camping out to "confront" the spirits and attempt to restore peace. They will be staying awake from Saturday evening (October 28) until dawn on Sunday, entertained by wardens' tales of the hauntings.'

Ultimately, however, the diabolical horsemen did not make a showing.

On May 1, 2004, Alec Williams was driving past the car-park that sits at the base of the Castle Ring when he was witness to a dark, hair-covered, man-like entity that lumbered across the road and into the attendant trees. Williams stated that the sighting lasted barely a few seconds, but that he *was* able to make out the shape of its monstrous form:

> 'It was about seven feet tall, with short, shiny, dark brown hair, large head and had eyes that glowed bright red.'

Interestingly, Williams stated that as he slowed his vehicle down, he witnessed something akin to a camera flash coming from the depths of the woods and heard a cry that he described as 'someone going "Hoooooo."' The beast did not resurface, and a shell-shocked Williams was forced to continue on his journey, wondering what on Earth – or possibly even off it – had just taken place.

Just over one year later, on June 8, 2005 to be absolutely specific, in an article titled *Hunt For Dark Forces at Chase Monument*, *Chase Post* writer Sarah Taylor reported that 'paranormal investigators are set to swoop on one of the area's oldest monuments to find out what dark forces lie beneath it'. As the newspaper noted, 'a team of real-life ghost-busters' had determined that the area of Gentleshaw that surrounds Castle Ring lay upon a 'psychic fault'.

Indeed, the whole area surrounding Castle Ring has been a hotbed of unusual activity for years – and not all of it revolves around weird beasts.

For example, commenting on the high-strangeness at the Ring, Sue Penton – of *Paranormal Awakening*, a group affiliated to the *Association for Scientific Study of Anomalous Phenomena* – said: 'There have been reports of strange music being heard up there. It is such a high place, there have been lots of UFO sightings there, too.'

This was amply echoed by Graham Allen, who at the time was the head of the Etchinghill, Rugeley-based *Staffordshire UFO Group*, and who had taken over the reins from the group's founder, Irene Bott, several years earlier:

> 'Obviously, Castle Ring is the highest point on the Chase which makes it a good place for UFO spotting. There have been numerous incidents of UFOs, which could be because you are more likely to see something from a high point.'

Allen elaborated that:

> 'There have been reports of something landing there in the 1960s. From a research point of view there are a high number of reports around ancient sites. One argument could be that ancient sites have been located there because of the incidents of UFOs and natural phenomenon. There could be locations where there could be magnetic influences in the ground which have been attributed to earth lights.'

Moreover, relatively close to the Castle Ring is an old, disused windmill, which, it is widely believed and accepted by local historians, was constructed upon the now-crumbled remains of ancient, pagan burial ground. Ghosts of the miller's children, who local legend says suffocated in a flour-silo, are said to haunt the mill to this day, and the folklore of the area tells of a strange black figure that appeared just before the tragedy. Could this perhaps have been the same dark figure which Alec Williams saw near the Castle Ring in 2004? Equally as strange are the reports from the village of Cannock Wood – from which the Castle Ring lies in a north-west direction – of a ghostly nun that has been seen in the vicinity of an ancient well.

In September 2005, the local media reported that the aforementioned *Paranormal Awakening* investigation group had recently completed a nighttime investigation of the Castle Ring in an attempt to try and chronicle the strange activity that had been reported there for years.

A spokesperson for the group told the newspaper:

> 'The Cannock Chase local authorities were kind enough to give permission for *PA* to conduct its research. Indeed, we are extremely grateful to them for being so open-minded as to allow us to conduct our research at this historical and most important monument. The group's results are stunning and have created yet more questions than we have answers. We appear to have obtained a very strange mix of UFO and genuine paranormal activity.'

Midway through February 2006, the *Chase Post* elaborated:

> 'A paranormal investigations group say they have evidence of strange, dancing lights and ghostly figures at the area's most ancient monument.' On one tape, said the *Post*, one of the group's members is heard to exclaim: 'Tell me that isn't a big black shape walking towards me.' The *Post* added that: 'A mystery make voice responds, "There is!"'

Of course, it should not be forgotten that large, dark shapes and strange lights were both staple parts of Alec Williams' 2004 sighting near the Castle Ring, too...

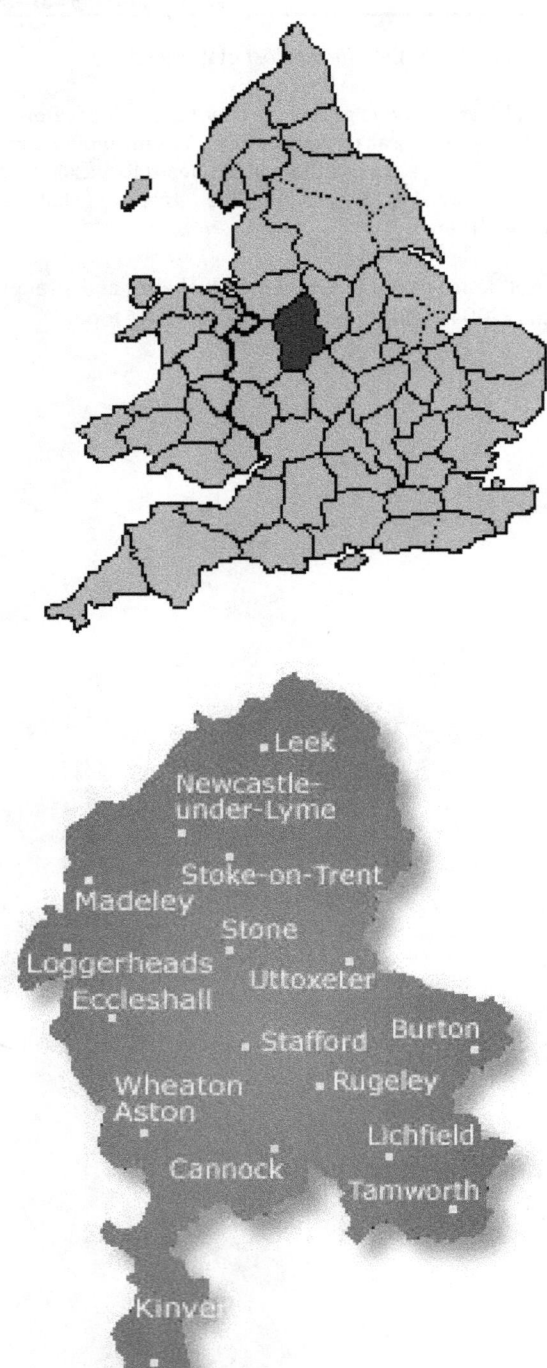

Chapter IX
Where the Bigfoot Lurks
'...The best way I can describe them to you is like a hairy troll...'

Throughout the annals of history, folklore and mythology worldwide encounters with hairy, man-like entities - such as Bigfoot and the Abominable Snowman - literally abound. Let us start with the monster of the frozen Himalayas.

The term Abominable Snowman was not actually coined until 1921, the same year in which Lieutenant-Colonel Charles Howard-Bury led the joint Alpine Club and Royal Geographical Society Everest Reconnaissance Expedition, the details of which he carefully and scrupulously chronicled in 1921 in *Mount Everest, The Reconnaissance*. In his writings, Howard-Bury included an account of crossing the Lhakpa-la at 21,000 feet when he stumbled upon footprints that he believed 'were probably caused by a large loping grey wolf, which in the soft snow formed double tracks rather like a those of a bare-footed man'.

However, Howard-Bury said that his Sherpa guides opined that the tracks were actually of 'The Wild Man of the Snows', to which they gave the name 'Metoh-kangmi' – 'Metoh' which translates as 'man-bear', and 'Kang-mi', which means 'snowman'.

And then there are the Almas, or Almasty of Central Asia, Russia, and the Caucasus – human-like bipedal animals, between five and six-and-a-half-feet tall, with bodies covered with reddish-brown hair, and anthropomorphic facial features that include a pronounced brow and a very flat nose.

Moving on to Australia, there is the legend of the Yowie. The mid-to-late 19th Century saw a plethora of encounters with, and reports of, such creatures, the majority of which involved large, gorilla-like creatures (albeit usually wholly bipedal), which lived in remote mountainous or forested regions. Reports have continued to surface right up until the present day, with the trail of evidence following the pattern familiar to most unidentified hominids around the world – i.e., eyewitness accounts, mysterious footprints of somewhat-disputed origin, and an unfortunate lack of conclusive proof one way or the other.

And then there is the most famous of all man-beasts: the Bigfoot of the vast forests of the United States of America, of course.

Sasquatch supporters Grover Krantz and Geoffrey Bourne believe that the creatures known to millions as Bigfoot may very well be a relict population of Gigantopithecus: a giant ape that roamed China, Tibet and elsewhere before its presumed extinction hundreds of thousands of years ago. Bourne states that since most Gigantopithecus fossil remnants have been found in China - and, as a result of the fact that many species of animals migrated across the Bering land bridge eons ago - it is not at all unreasonable to assume or posit that Gigantopithecus might have as well, and, therefore might not be so extinct, after all.

Nevertheless, despite the merits of the argument, the Gigantopithecus hypothesis is generally considered entirely speculative for the mainstream scientific and zoological communities. Gigantopithecus fossils, for example, have never been found in the Americas. And, taking into consideration the admitted fact that – thus far - the only recovered fossils of the mighty beast are of mandibles and teeth, there is actually some very deep uncertainty about the exact nature of Gigantopithecus' locomotion and gait.

Krantz, for example, has argued, based upon his extrapolation of the shape of its mandible, that Gigantopithecus blacki could very well have been bipedal in nature. However, the relevant part of the mandible is not present in any of the fossils found and examined thus far. In reality, the mainstream view is that Gigantopithecus was a quadruped rather than a bipedal entity, and it has been argued that the enormous mass of Gigantopithecus would have made it very difficult indeed for the creature to adopt a bipedal gait to any meaningful degree.

Matt Cartmill offers another stumbling-block associated with the Gigantopithecus theory: 'The trouble with this account is that Gigantopithecus was not a hominine and maybe not even a crown-group hominoid; yet the physical evidence implies that Bigfoot is an upright biped with buttocks and a long, stout, permanently adducted hallux. These are hominine autapomorphies, not found in other mammals or other bipeds. It seems unlikely that Gigantopithecus would have evolved these uniquely hominine traits in parallel.'

Bernard G. Campbellin says of this same controversy: 'That Gigantopithecus is in fact extinct has been questioned by those who believe it survives as the Yeti of the Himalayas and the Sasquatch of the north-west American coast. But the evidence for these creatures is not convincing.'

And so the controversy surrounding the giant, hairy beast-men (and women, of course) of Russia, the United States, China, Australia and elsewhere continues. But: what of the British Isles? And: what about Staffordshire?

As incredible and wholly unbelievable as it may well seem at first glance, sightings of large, hairy, ape-like animals have abounded throughout Staffordshire for more than 100 years. It is one thing to claim a sighting of such a lumbering ape-man-style beast deep within the huge forests of the United States' Pacific Northwest, or high upon the snow-dominated slopes of the harsh Himalayas. But it is altogether quite another matter to claim an encounter with a Bigfoot-like animal right in the middle of rural Staffordshire. Countless witnesses have incredibly claimed precisely that, however. Without any shadow of doubt, the most famous of all Staffordshire man-beasts is the Man-Monkey: a hairy, ape-like beast with bright, shining eyes that in January 1879 burst forth out of the dense woods that surround Bridge 39 on the Shropshire Union Canal, as well as within the nearby Staffordshire village of Ranton. Since that famous date – as I noted in my 2007 full-length book on this very subject, *Man-Monkey: In Search of the British Bigfoot* – this particular man-beast has now been seen on close to twenty occasions, at least: usually still preferring to hang out in and around its original haunts, where it seems to take a great deal of unhealthy pleasure in scaring half-to-death anyone that dares to cross its unholy and vile path.

Since the long, winding and convoluted tale of the Man-Monkey is related at length in the aforementioned book, and as additional, new data on the puzzle was published in my 2012 book, *Wildman! A Guide to British Man-Beasts*, I have chosen to focus in these pages on other, lesser-known stories of a hairy and beastly nature that emanate from Staffordshire.

I am not at all sure that the following case is directly linked to the controversies surrounding Bigfoot or unknown apes in Staffordshire. Frankly, however, I can think of nowhere else in this book to place the story! But, as it *does* involve the sighting of unidentified, hairy humanoids, I conclude that here is just about as good a place as any to relate the admittedly-odd facts.

While there can be little doubt at all that the vast majority of the reports that appear within the pages of this book are both controversial and sensational, the following is certainly one of the weirdest of all. Understandably, given the bizarre set of circumstances, the family in question is highly reluctant to speak out publicly, aside from revealing the basic facts of the case.

Doubtless, the arch-sceptic or debunker would conclude that this determination on the part of the family not to discuss their case on the record is indicative of the fact that I have been the victim of an audacious hoax or a practical joke. And, while such a scenario can never be outright dismissed, of course, having now spoken to the husband and wife concerned on several occasions and twice now in-person, I do not doubt the veracity of the case or of their honesty and sincerity.

It was in the early hours of a winter's morning in 1975 when Barry and Elaine, a married couple then in their late twenties and with two small children, were driving towards their then-Slittingmill home, after attending a Christmas party in the Staffordshire town of Penkridge. As the pair headed towards the village, their car's engine began to splutter and, to their consternation and concern, completely died. Having managed to carefully coast the car to the side of the road, Barry proceeded to quickly open the bonnet and take a look at the engine – 'even though I'm mainly useless at mechanical stuff', he states.

There did not appear to be any loose-wires, the radiator was certainly not over-heated, and a check of the car's fuses did not provide any indication of what the problem might be. And so, as the family was less than half-a-mile or so from home, Barry made a decision, as he explained: 'We had a picnic blanket in the boot of the car and I got it out, got back into the car and said to Elaine something like: "Let's cover up you and the kids with the blanket." They were in the back sleeping and [were] only four and six at the time. So I said to [Elaine]: "You stay with them, and I'll run back home and get your car, pick the three of you up, and then we'll leave my car here, and we can get a garage out to look at it tomorrow."'

At that point, however, their plans were thrown into complete and utter disarray. According to Barry, Elaine let out a loud scream, terrified by the sight of a small figure that ran across the road in front of them at a high rate of speed.

She explains:

> 'I just about saw it at the last second, and then another one followed it, and then a third one. The best way I can describe them to you is like a hairy troll or something like that. We had some moonlight and they were like little men, but with hunchbacks and big, hooked noses and not a stitch on them at all. Not a stitch, at all; just hair all over them. I'd say they were all four-feet-tallish, and when the third one had crossed by us, you could see them at the edge of the trees – sort of like they were watching us, like they were wary or something anyway.'

At that point things became very hazy indeed, says Barry:

> 'We both know from memory that they came forwards, towards us, very slowly to us, and I've thought since that they were interested in us or wanted to know who we were. They came very slowly, and it was a bit like we were being hunted, to me. Elaine was hysterical; and with the kids with us, I wasn't far-off, either.

> 'But that's all we remember. The next, it's all gone: nothing. Neither of us remembers seeing them go; and the next thing it was about two o'clock and the car started fine, then. It felt like something had happened to us that I couldn't quite put my finger on, you know? But the memory thing is the biggest problem; even now. What was it? I *did* have a dream later about them surrounding the car; but that's it, really. But they were there and we did see them, right up by the *Stone House* [Author's Note: A reference to a large, old house that sits on the edge of the village of Slittingmill and that dates back to 1584].'

Barry states that to this day both he and Elaine still feel very uneasy about the loss of memory that they both experienced, but he is keen to affirm that:

> '*I* know, and *we* know, that we both saw them. The kids don't remember a thing, thankfully. They were horrible little things: trolls, goblins, something. But they *were* there and they *were* real.'

Neither Barry nor Elaine have ever experienced any further such incidents or encounters with the unknown, but they have never forgotten those disturbing events deep in the heart of the Cannock Chase on a chilly, winter night all those years ago.

Late one evening in 1986, Mick Dodds and his wife were driving by the ancient and ruined

Chartley Castle that, today, overlooks the A518 road at Stowe-by-Chartley. Built on land that had come into the possession of the Earls of Chester as far back as the end of the 11th Century, Chartley Castle was a stone motte-and-bailey fortress founded in the thirteenth century by Ranulph Blundeville, the then Earl of Chester. Supported by the motte are, today, the still-standing remains of a rare cylindrical keep, with the inner bailey curtain wall still strongly flanked by two huge half-round towers, a gate-house, and an angle-tower. A strong counter-scarp bank and cross-ditch divides the inner and outer baileys, with another ditch and bank encasing the whole castle. Notably, Chartley Castle is to where – on Christmas Eve, 1585 – Mary, Queen of Scots was taken before being moved to Fotheringay for execution on February 8, 1586.

And with that bit of brief background information out of the way, let us now return to the story of Mick Dodds. He asserts that as they negotiated the dark road, his wife suddenly screamed at the sight of what looked like a large chimpanzee that bounded across the road, and right in front of their car. Half way across the road, the 'chimpanzee' stopped suddenly, looked directly at the terrified husband and wife and, to their utter horror and consternation, charged their vehicle – but, at the last moment, backed away from actually causing any structural damage to the car, or physical harm to the fear-stricken pair.

Dodds said that in his overwhelming panic to quickly put the vehicle into reverse gear, he stalled its engine; and then, even worse still, ended up completely flooding it as he tried to re-start the car. As an inevitable result, the Dodds were briefly stranded in the road with a hairy monstrosity looming wildly in front of them. For about twenty seconds the beast stared at both husband and wife, and on two other occasions again headed for their vehicle at full speed, 'like it was going to attack', before finally bounding off to the left. It did not return again to terrorise the couple.

In 1995, Jackie Houghton was living in Cannock and working as a waitress in the nearby town of Stafford. On February 18 of that year, and at around 1.00 a.m., she had been driving across the Cannock Chase, and along the main road that links the towns of Rugeley and Cannock, after her shift at the restaurant was finally over. As she approached the turning for the village of Slittingmill (about which, intriguingly, we have already heard quite a lot indeed), however, she was suddenly forced to violently swerve the car and only narrowly avoided collision with a large, shambling creature that stepped out into the road at a distance of about two hundred yards from her.

Considering that she was travelling at high speed, said Jackie, it was an absolute wonder that she didn't hit the thing. The encounter lasted just a few seconds, but she had caught sight of the animal in the headlights of her vehicle and was certain that it was man-like and tall, very hairy, and with two brightly-glowing red eyes. In an instant, said Jackie, the beast vanished into the cover of the surrounding trees, leaving her distinctly shaken and highly stirred.

Gavin Addis claims a very sensational encounter with a Bigfoot-like entity at one of the Chase's most famous attractions: the Glacial Boulder. Made out of granite, the boulder is both large and impressive. It is also made curious by virtue of the fact that there are no natural granite outcrops anywhere in the area. Indeed, the nearest rock of this type can be found in the Lake District, more than 120 miles to the north, and on Dartmoor, Devonshire, no less than 165 miles to the southwest. The boulder, however, has been matched conclusively to a rocky outcrop at Cniffel in Dumfries & Galloway, which is over 170 miles from the Chase in the Southern Uplands of Scotland. At some point during the last Ice Age, the boulder must have been carried by the great glaciers down the country and to its present location on the Cannock Chase.

According to Gavin, on a winter's night in 1997, he and his girlfriend were parked in his car near the boulder, doing what courting couples all across the world do late at night, when his girlfriend let out a loud and hysterical scream: standing atop the boulder itself was nothing less than a large hairy man, waving his arms wildly at the starlit sky. With good presence of mind, Gavin jumped into the front-seat of the car and floored the accelerator. Tires spun, dirt flew into the air, and the car shot away at high speed, but not before the creature succeeded in jumping onto the bonnet of his car. For five harrowing minutes, it hung on, before finally being thrown to the ground. Gavin looked in his rear-view mirror and could see the creature already back on its feet and running in the direction of the woods.

It must be said that several people who have met Gavin Addis are convinced that his sensational tale is simply that: namely, the wild and unbelievable ravings of a slightly pathetic fantasist and nothing more. For his part, and perhaps even to his credit, Gavin totally understands and appreciates this reaction; but he has been very careful to point out in response that he has absolutely *nothing* to gain – and, arguably, absolutely *everything* to lose – by fabricating such a strange and unlikely story. And, in that respect, he is not wrong at all: saying that you have seen Bigfoot on the Cannock Chase (and that the infamous man-beast succeeded in jumping onto the bonnet of your car, no less!) is, unfortunately, unlikely to result in anything other than the rolling of eyes and loud hoots of derision.

Submitted to the website of the United States-based group *Gulf Coast Bigfoot Research Organisation* by researcher Bobby Hamilton is the following account from a confidential source, who reported seeing a strange, man-beast on the Cannock Chase in September 1998. With three friends, the source was journeying by car along the A34 road between Stafford and Cannock. It was about 12.30 a.m., and in the area of Cannock Wood, when all four simultaneously noticed something distinctly strange just off the side of the road.

In the words of the chief source of the account:

> 'It was a star-filled night, clear, but dark and we were all in the car driving home, happily chatting and joking. Suddenly we all fell dead serious, the people in the back sat forward and we all pointed to the same shape. It was a tall man-like figure, sort of crouching forward. As we passed, it turned and looked straight at us. In my own words I would describe it as around 6 feet 8 inches tall, legs thicker than two of mine, very strong looking and with a darkish, blacky-brown coat. I just could not explain it and I still get goose-bumps thinking of it.'

The informant advised Hamilton that: 'No one would be out there on the night playing about;

it was very cold that night…I can identify deer in thick bush; this was open. I hope this is of some help; it is the absolute truth to the best of my memory. Why make up something like that?' To which I can only briefly add: why, indeed?

The next Bigfoot report of any real, meaningful substance surfaced in January 2003, when Peter Rhodes of the *Express and Star* newspaper wrote an article pertaining to an extraordinary encounter on the Cannock Chase that eerily paralleled the controversial tale of Gavin Addis of some five years previously. Rhodes reported under the graphic and memorable headline of 'Night Terror with a British Bigfoot' that:

> 'Whatever it was, it scared the living daylights out of Craig Blackmore. His mother Val says: "I have never seen Craig like that before. He came home shaking, absolutely petrified and white, as though he'd seen a ghost."'

What Craig – and a friend of his named Jo – had actually seen was not a ghost but a huge, ape-like creature at the side of the road on Levedale Lane between Stafford and Penkridge. Craig told Peter Rhodes that:

> 'I was driving my Fiesta down the road towards Penkridge and as we approached a house, the security light came on. I saw something in the corner of my eye. It was coming towards the car, running very fast. It wasn't a dog or a deer. It was running like a human would run, but it was really hairy and dark. It came level and jumped at the car but just missed. My friend, Jo, turned round and said it was huge and had run through the hedge and across the field. I turned the car around but there was no sign of it.'

Craig's mother added that:

> 'I thought maybe Craig had been drinking, or perhaps someone had spiked a drink. But that hadn't happened. He is a very truthful boy. He would not say something had happened if it hadn't. And anyway, his friend, Jo, was in the same state of shock.'

Peter Rhodes noted:

> 'Although the event had been terrifying, Craig, a 19-year-old HGV mechanic, did not report it to the police. He told a few friends ("they all laughed") and tried to forget the experience.'

Rhodes also spoke with British-based Bigfoot investigator Geoff Lincoln, who told the *Express and Star* feature-writer that: 'Bigfoot in Britain is an odd subject and very often the target of ridicule. But sightings are taking place and I am currently looking into two other reports in 2002, one in Northumberland and another in Lancashire.'

To Craig Blackmore, Lincoln offered a simple message (and one that Peter Rhodes said was 'worthy of *The X-Files*'): 'You are not alone.'

Perhaps those who are highly suspicious of the account of Gavin Addis should now consider reevaluating their views somewhat, taking into consideration the startling similarities present in the encounter of Craig Blackmore and his friend, Jo.

In 2003, and following in the wake of the publication of the encounter of Craig Blackmore, none other than BBC Television personality, weatherman, and star of *The Morning Show*, Ian McCaskill became embroiled in the mystery of the Cannock Chase Bigfoot when, with cryptozoologists and Centre for Fortean Zoology-stalwarts Jonathan Downes and Richard Freeman, he headed to the area in hot-pursuit of the mysterious creature himself.

The BBC was not surprisingly full of good-natured humour and banter as it reported on McCaskill's adventures on, and romps in and around, the Chase:

> 'Apparently he's hairy, giant and ape-like. And not at all the sort of person you want to bump into in a deserted place, on a dark night. Unfortunately, "Bigfoot", or the "Yeti" as it's become known, is out and about, on the prowl, and could be coming to a place near you. Never fear, for the *Morning Show's* gallant Ian McCaskill was here. In the first week of February, he went off to Cannock Chase in Staffordshire to hunt down the eight-foot Yeti spotted recently in the area.'

The BBC continued:

> 'Ian joined monster hunters Jon Downes and Richard Freeman, from the Centre for Fortean Zoology based in Exeter. Richard, a qualified zoo keeper and centre founder Jon describe themselves as "Britain's foremost professional monster hunters". They were following up the recent sighting on the side of the road near Stafford. It was only miles from another "Bigfoot" appearance four years earlier. Although they have been mocked, a UK website devoted to Bigfoot research contains many reports, including yet another Staffordshire sighting. Lots of sightings occur near telecom towers, and one of the theories is that apparitions are caused by radiation surges from these towers.'

Situated near the picturesque, ancient Staffordshire hamlet of Milford, Shugborough Hall is both a large and renowned country house that serves as the ancestral home of the Earls of the city of Lichfield; and its spacious grounds are connected to the nearby village of Great Haywood by the Essex Bridge, which was built during the Middle-Ages. Around 1750, the hall was greatly enlarged, and then yet again at the beginning of the 1800s. Today, Shugborough Hall is open to the general public and boasts a working farm museum that dates back to 1805, and which is complete with a watermill, kitchens and a dairy.

Interestingly, the grounds of Shugborough Hall are also home to something known as the Shepherd's Monument, upon which can be found a very strange inscription, and one which many students of the puzzle believe may very well contain a top-secret code that identifies the alleged resting place of none other than the legendary Holy Grail itself. The Shepherd's Monument is not the only such construction of note on the grounds of the sprawling hall: The Tower of Winds; the Cat's Monument; and the Doric Temple also have pride of place at the historic hall.

Not only that: the woods and fields that surround Shugborough Hall are said to be the domain of nothing less than a diabolical, hairy wild man, and one that has been reportedly seen as late as 2004. In this particular case, the witness had impeccable credentials. She was a policewoman who, while on duty with a colleague on the night in question, was routinely patrolling the roads that run through the woods that surround the huge Shugborough estate.

It was not long before midnight when both she and her partner were shocked by the sight of a strange beast that charged across the road, only a short distance in front of them, and that proceeded to head towards the expansive fields that dominate Shugborough Hall.

In a personal interview with me that was undertaken in the summer of 2006, when I and my first wife Dana returned to England for several months, the officer described the animal as being human-like in shape, an incredible eight-and-a-half-feet in height, covered in dark hair, but looking practically emaciated in terms of its physique. Her colleague, who was driving, quickly slammed on the brakes and brought the car to a screeching halt in the middle of the moonlit road. The shocked pair looked at each other in understandable silence for a moment; but then, having regained their composure, elected to do absolutely nothing at all beyond continue with their nighttime patrol of the area.

No-one would believe their story, the officer told me, in somewhat frustrated tones. They would likely receive nothing more than ridicule from their colleagues if they dared speak a word of the night's events; and, in addition to that, what purpose would it serve to alert the staff at Shugborough Hall to the fact that a creature akin to Bigfoot was prowling around the area after night had fallen upon the area? No purpose at all, they quickly concluded.

Of course, this particular case begs a very intriguing question: how many *other* people may have seen the monster of Shugborough Hall, but who have also elected to say nothing, for fear of similar potential ridicule and endless jokes at their own, personal expense?

There is one, final point worth noting. And it's a highly significant one, too. Shug – as in Shugborough - is an ancient English term derived from an even older Anglo-Saxon word, Scucca, which means demon. Shugborough: the borough of the demon; how very appropriate, one might very justifiably conclude.

In the warm summer of 2005, Tom, a resident of the town of Bloxwich, spent several memorable nights camping out with two friends deep in the woods near the Cannock Chase's German Cemetery. Having made their camp, the friends headed off into town to purchase various items that they would need for their time in the great outdoors. On their return, however, they were angered – and disturbed, too – to find that 'something' strange had paid a visit to their campsite while they had been gone. Charcoal bags, clothes and much more had all been 'flung around' the campsite in a very haphazard fashion, as if by some form of irate, wild animal.

The vexed trio duly set to work and cleaned up the area, then got on with the day's activities. Around 2.00 a.m. on the following morning, however, and after they had retired to their respective beds, all three were jolted awake by a hideous, animalistic scream that emanated from within the woods. It was quite unlike anything Tom and his friends had ever heard before, he carefully assured me. Somewhat concerned, on the following day they once again headed into town for provisions. But, on their return, they were faced with an even greater puzzle: their possessions were as they had left them, but now their tent was gone.

Of course, some might say that the strange scream was merely that of an everyday fox, but Tom is adamant that it was something far, far stranger indeed. He also reiterates the fact that the group was camped deep in the woods; and, therefore, the chances of anyone stumbling across their camp were virtually zero. Also: as Tom rightly notes, if people *were* responsible for the strange and unsettling activity at the campsite, then why were the group's valuables not stolen along with their tent? And, moreover, what would lead someone – or maybe something - to haphazardly fling their possessions around their tent on the previous morning? Clearly, five years later, the answers to those questions are now unlikely to ever be forthcoming. However, Tom is certain in his mind that *something* was out in the woods on those dark nights, watching, screaming and taking a keen and unsettling interest in the activities of both him and his two friends...

Chapter X
Staffordshire Bigfoot Mania
'...Reports were circulating today that a strange beast is roaming Cannock Chase...'

In the latter part of 2005, I penned an article titled *Bigfoot in Britain* that was published in the U.S.-based *Fate* magazine in January 2006. It was this article that subsequently - albeit wholly inadvertently, I have to say - set off a strange and ever-escalating chain of events that saw the town of Cannock and the surrounding Cannock Chase plunged into a Bigfoot craze that shows no signs of stopping, even several years further down the line. Quite literally, the whole area seemingly went Sasquatch mad, with local media outlets quickly jumping on the controversy at every given opportunity. Indeed, *Chase Post* Editor Mike Lockley would himself concede that it was my article that propelled the controversy of 2006 to stratospheric levels, when he said: 'Bigfoot fever first struck in our area when respected "*X-Files*" reporter Nick Redfern started investigating the existence of a strange creature after a spate of sightings over the Chase.'

On February 14, in an article titled 'Is There Truth In Bigfoot Sightings?' the *Birmingham Mail* stated that:

> 'Reports were circulating today that a strange beast is roaming Cannock Chase – and now paranormal investigators are investigating.' And as the *Mail* also recorded in its pages: '...all of the locals who spotted the mystery beast give the same description: a giant, hairy creature with blazing red eyes.'

And, thus began a truly surreal period in which all things man-like, large and hairy were forever dominating for the folk of the Cannock Chase.

Indeed, recognising that the controversy was very likely to provoke both deep interest and high entertainment in equal entertaining measures, the *Chase Post's* editor Mike Lockley noted in the February 9 edition of the newspaper that: 'Anyone who can provide me with a picture of Bigfoot will get a free Indian meal.'

And, of course, photographs – all carefully crafted with the benefit of Adobe Photoshop - poured in to the offices of the 'Post', most of which were accompanied by humour-filled accounts, including that of Andrew Soltysik, who submitted the following, along with a photograph of a giant, silverback gorilla superimposed on the Chase:

> 'By chance I was walking my dog, Rusty, up by the raceway and the poor fellow came racing towards me, tail between his legs and shaking with fright. Moments later, over the brow of the hill, I saw what I can only describe as a beast. I was able to take a shot of him before I myself turned and ran.'

Commenting on his *King Kong*-style photograph, Soltysik added:

> 'You can tell from the scale of things that he must be far in excess of the 6ft 8ins mentioned in your story. To be honest, I was too frightened to hang around, his exact height wasn't too important at the time! I think that in his quest for survival on Cannock Chase he must have picked up a discarded bag of chips wrapped in the *Chase Post* and noticed the story within. Perhaps he's a little upset that his cover has now been blown. Whatever his reasons, he didn't seem in a very joyous mood. I returned about half an hour later but could find no trace of him.'

Soltysik closed thus:

> 'Over the weekend I intend to hunt him out. I know of a few caves, hideouts and tree camps where he may be setting up home. Wish me luck!'

Mike Lockley again made his generous offer of a 'free Indian meal' for anyone who could provide a 'genuine picture of Bigfoot' on the Chase. If that were to fail, however, Lockley added that: '…we'll select the one that gives us the best laugh.' Cue a stampede of even more Photoshop-created imagery that provided very welcome light relief from the stranger, and far more disturbing, reports of seemingly genuine Bigfoot-style encounters on the Cannock Chase that we have already examined and chronicled.

It is very intriguing to note that when reports of Bigfoot on the Cannock Chase first began to circulate widely within the Staffordshire media in 2006, they were actually not dismissed out of hand by local wildlife experts. Granted, those same experts were most certainly *not* prepared to accept that the legendary ape-man had made a home for himself on the Cannock Chase, but they were willing to accept that people were at least seeing *something*. One of those was Derek Crawley, the Chairman of the *Staffordshire Mammal Group*. Crawley said on February 22 that:

> 'From the reports it would seem too big to be an escaped ape. A lot of people seem to be saying it has red, staring eyes. It may be a big cat in a tree. The person who saw it may have been confused and mistaken the branches for arms.'

At the same time, Chris Mullins of *Beastwatch UK* called for a further investigation, and had the following to say in response to the theory of Derek Crawley:

> 'There is definitely something weird going on and the theory that some of them could be big cats in trees is a genuinely interesting one.'

And, of course, the jokes continued to pour into the offices of the local newspapers. Said one 'M. Willis' of the town of Hednesford:

> 'Some weeks ago I was walking in the woods of Beaudesert Old Park, when a large hairy being crossed my path. As it passed it turned and glared right at me. I never forgot those eyes, and some days later I opened the *Chase Post* and the truth hit me. The creature I had seen was none other than Big Dave [a columnist] from the *Post*. Was he on an undercover story or has he got a weird hobby?'

Similarly, the newspaper stated with good humour that:

> 'Cannock man Mr. Hanysz even sent us a snap of what he believes could be the beast. Whether it's our furry friend or one of the CHAVs at chucking out time, the jury is out again.'

One week later Chris Mullins had yet more fuel to add to the already-escalating fire:

> 'Theory has it that a Bigfoot couldn't possibly exist in our country due to the small area of rough country. However, I believe they could be creatures that may well dwell underground. According to reports that I've taken myself, all sightings have been at night or in very poor light – this may be due to their eyes being sensitive to light.'

Mullins continued:

> 'As far as their eyes glowing red are concerned, I cannot possibly put a logical explanation to this. I must admit I don't follow the big cat theory as I've not only studied but even witnessed big cat in the UK. Their eyes tend to reflect a golden yellow to almost white and I have never heard of a big cat's eyes reflecting red as yet.'

The strange saga of the Staffordshire Bigfoot reached what were truly surreal proportions when, on March 23, 2006, the *Chase Post* included a story that easily eclipsed the *Sun's* infamous *Freddy Starr Ate My Hamster* episode of the 1980s, when it reported in wholly serious tones that *Bigfoot Almost Caused Me To Lose My Baby*!

Stated the 'Post':

> 'Police chiefs have hit out at the dangers posed by the spoof "Bigfoot" craze after a teenager almost lost her baby when a joker clad in a gorilla suit jumped in front of her car. And the concerns have been echoed by a leading councilor and conservationist, who fears the "irresponsible idiots" are causing harm to wildlife as well as people.' The story was a bizarre one, to say the very least.

The odd incident in question had apparently occurred around 11.00 p.m. the previous weekend, in a dimly-lit woodland road in the village of Brocton, as the 19-year-old girl and her family drove home after they had all dined in the pleasant, little, nearby locale of Milford. The girl said as she recounted her story that: 'We noticed a BMW parked in the road. Suddenly it flashed its lights. Just then, out of nowhere, this person dressed in a gorilla suit jumped out in front of our car, flailing their arms like mad. Then they started running at the car like mad. It was terrifying.

> 'Looking back it was obviously a fake suit, but late at night, in an isolated area like that, it was a very scary experience. I broad daylight, I suppose it could be quite funny, but this was 11 o'clock at night with no-one around. It's very lonely there. If that had been someone with a heart condition, they could've had a heart attack. I screamed so loud. It was a real scare. It left me with fears that the trauma of it could have fatally harmed my baby.'

The girl's father firmly, concisely and quite understandably told the *Chase Post* in irate tones that: 'If I'd have caught the idiots, I'd have pasted them.'

And the local police were hardly in a mood for much laughter, either. As evidence of this, a spokesperson for Staffordshire Police HQ said: 'We take it very seriously because it may result in a Public Order Offence. The person [in the gorilla suit] may very well be in high spirits, but this would be viewed as a criminal offence.'

Local councillor John Burnett was determined to have his say on the monstrous matter, too: 'This is the behaviour of an irresponsible idiot. At this time of year, there are all manner of ground-nesting birds in that area; the partridge, the pheasant, woodlarks, skylarks – many rare birds whose habitat and nesting could be destroyed by this kind of activity.'

While the wild stories that appeared in the *Chase Post* prompted intrigue, laughter, and general interest for the people of the Chase, their American cousins seemed to be far less impressed or amused by the crazed antics that were afoot on what was practically a daily basis at the time.

The 'Post' noted that one contributor to a particular Bigfoot forum had stated with respect to the flood of photographs that had been sent to its offices:

> 'There are interesting aspects of the whole Cannock Chase mystery that could be investigated and that would prove genuinely interesting. But clearly clicking about on Photoshop for five minutes is easier than getting out and researching a story. The sad fact is that all this will do is make people less likely to come forward for fear of being ridiculed so the whole story grinds to a halt and the actual solution to the mystery remains elusive.'

And then there was the following, 'equally damning' comment, as the *Post* went on to describe it:

> 'I keep waiting to see how low my opinion of the coverage of the Cannock Chase Bigfoot story can get and every time I think they can't get any lower they

somehow manage it. The stories are accompanied by the god awful collection of Photoshopped snaps with pictures of gorillas and Chewbacca pasted into a forest scene.'

The staff of the *Chase Post* was completely unmoved by this comment, however, and proudly said to just about one and all:

> 'We at the *Chase Post* are not so quick to condemn. We believe people should be given the right to send us in pictures they believe show their encounters with the mystic beast. To all those doubters, we ask you this: If you've never seen Bigfoot, how can you tell a person he does not resemble Chewbacca or King Kong? We rest our case.'

Of course, this only infuriated those with no sense of humour in the Bigfoot-hunting community even more (and, believe me, there are a *lot* of them); much to the satisfaction of the *Chase Post*, one strongly suspects.

Etching Hill, that towers over the Staffordshire town of Rugeley, was the site of an encounter with a chimpanzee-like animal that was seen early on the morning of May 7, 2009 by a man driving to work, and who, in a brief email sent to me eleven days later, described the animal as being: '…about four-feet-tall, with a big piece of wood in his right hand. He was black and hairy all over and shot off as I drove past. It was definitely some sort of biggish monkey though.' It may not be entirely coincidental that Etching Hill has an appropriately intriguing history.

An article that appeared in the *Lichfield Post* on January 15, 2010 and titled 'The Little-Known History of the Hill', makes that point amply clear:

> 'Towering over Rugeley, Etching Hill stands proudly, just as it has for hundreds of years, overlooking the town and a central part of the community. But, the historic site which lies in an area of outstanding natural beauty and is listed as one of Staffordshire's RIGS (Regionally Important Geological and Geomorphological Sites). Etching Hill has had many uses over the years and has not always been the picturesque beauty spot it is today. In its time, the hill, which was formed during the ice age, has been home to armed forces, visitors from across the region as well as a race track but there are many people who know little of the importance of the site. Etching Hill stands 454 feet above sea-level on the edge of Cannock Chase and is capped with an outcrop of rock covered on its lower slopes to the north, south and west with gorse, bracken and heather and on the east by fields.
>
> 'When exactly horse racing began at Etching Hill is unclear; however, what is known is that infamous poisoner Dr. William Palmer was once a clerk at the site. Foot races were being run on a course at Etching Hill as early as in the reign of King Charles II when noblemen and gentry employed especially fleet footed men as "footmen" - men who ran before and after their masters' coaches. The course

was also present from 1820- 1856 and ran along the foot of the hill, along what is Mount Road, down towards Chaseley House out over towards Shooting Butts Farm and back down to the Hill. At the foot of Etching Hill, the old rifle range had its iron targets which were whitewashed and then painted with black paint.

'And scientists from Birmingham University excavated the site over a period of two months in 1974 after the discovery of two skulls in a cave at Bowers Farm opposite the bottom of Etching Hill. What they found was astonishing. They discovered 267 flints, among them, scrapers for the preparation of game, many microliths which were used as the tips or barbs of arrows and a single serrated flint knife. There was also evidence that the site had been used for knapping and tool manufacture. The cave site turned out to possess the third largest late Mesolithic flint assemblage known in the county.'

As with the various man-beasts seen at Staffordshire's Castle Ring, at the Glacial Boulder and at Chartley Castle, it seems that the creature of Etching Hill is similarly drawn to locations of a distinctly historic nature…

The site of Nigel Lea's Black Dog Encounter of 1972. Nick Redfern, 2006.
(see Chapter 1)

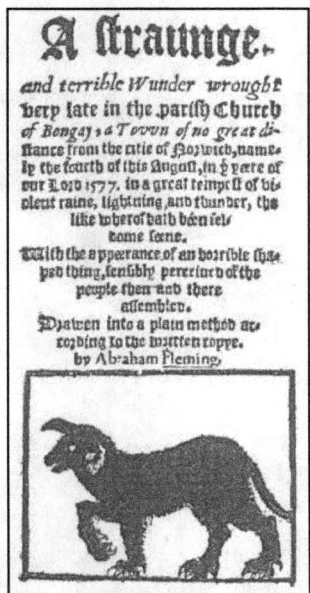

Beware of the Black Dog. Abraham Fleming, 1577.

Staffordshire's gruesome fish-people. Unknown, 1866. (see Chapter 2)

There's something in the water. John William Waterhouse, 1901.
(see Chapter 2)

The House of Commons was the scene of a historic debate on Britain's big-cats in 1998. Pugin & Rowlandson, 1808.
(see Chapter 3)

In 1996, a lynx was seen near the Staffordshire town of Penkridge. U.S. Fish & Wildlife Service. (see Chapter 5)

Staffordshire - the lair of exotic cats. Nick Redfern, 2009. (see Chapter 4)

Alien big cats - a staple part of Staffordshire. U.S. Fish and Wildlife Service, 2002.

Supernatural big cats in the woods. Lizars, date unknown.

Above: Castle Ring - the home of mysterious beasts. Nick Redfern, 2006.
Below: Bigfoot lurks at Castle Ring. Nick Redfern, 2006. (both Chapter 8)

Above: Castle Ring, where high-strangeness dominates. Nick Redfern, 2006.
(see Chapter 8)
Below: Numerous reports have been made of hairy wild men on the Cannock Chase. Nick Redfern, 1996. (see Chapter 9)

Wild men on the Cannock Chase. Jost Amman, 1589. (see Chapter 10)

Above: Roman View Pool, the rumoured home of a monstrous crocodile!
Nick Redfern, 2006.
Below: Sadly, the Roman View Pool crocodile turned out to be, in all probability, a common snapping turtle.
John Edwards Holbrook, (1842)
(both Chapter 11)

The big, bad wolf. Gustave Dore, 1883. (see Chapter 12)

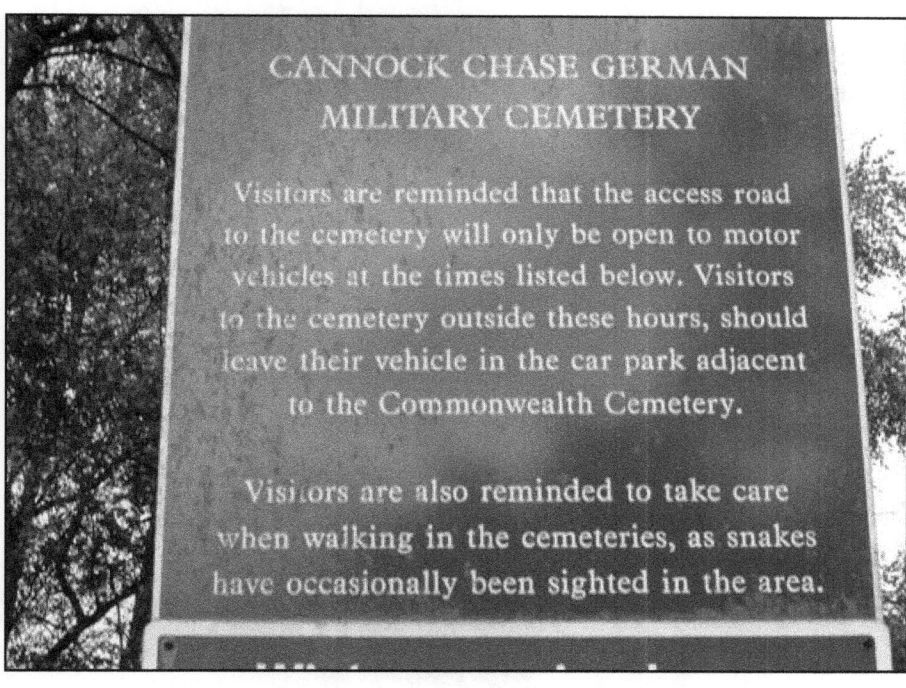

Above: The German Cemetery - the realm of the werwolf. Nick Redfern, 2006.
Below: A cemetery of terror. Nick Redfern, 2006.
(both Chapter 12)

Werewolves on the rampage. Unknown, 18th Century.
(see Chapter 12)

Above: For decades, the wallaby has made the wilds of Britain its home.
(CFZ, 2004)

Tusked thugs roam Staffordshire. Johannes Gehrts, 1901.
(see Chapter 14)

Above: The armadillo - a one-time resident of the city of Lichfield!
Friedrich Specht, 1927.
Below: In 1992, a pair of Coypu was encountered in the Staffordshire village of Alrewas. Friedrich Wilhelm
(Chapter 15)

Back in the 1930s, a monstrous, flying creature haunted Staffordshire's Bagot's Wood. Heinrich Harder, 1920. (see Chapter 16)

Chapter XI
There's something in the Water
'...My initial reaction was it was a crocodile or alligator and so I called the police...'

In the hot and humid summer of 2003, hysterical and pretty much near-unbelievable rumours wildly spread around the town of Cannock to the effect that a giant, marauding crocodile was on the loose in the area. Yes, really! Local police, the nation's press, and even representatives of the Royal Society for the Prevention of Cruelty to Animals (RSPCA) all wildly rushed to the scene of the action – which was the Roman View Pond, just outside of Cannock – and valiantly sought to determine the full facts about what, amongst the local media and the populace, was fast becoming known as the 'Cannock Nessie'.

As a result of all the burgeoning publicity that quickly surrounded the odd affair, a team from the Devon-based Centre for Fortean Zoology - which was led by CFZ Director and the editor and publisher of this very book, Jon Downes - headed off from Devon and made their convoy-like way to the little, innocuous pond where all of the animal-action was said to be taking place. And, since Jon Downes, arguably more than anyone else, was in a prime position to chronicle the facts and the expedition's highlights, it is over to Jon that I now happily turn you.

In Jon's very own words and style:

> 'The affair started with an e-mail message from Nick Redfern. He may be living in Texas now; living proof that one can take the boy out of the West Midlands. But, the fact that he still keeps a finger on the pulse of the event's of his hometown, prove that one may not be able to take the West Midlands out of the boy! It was a story from the *Wolverhampton Express and Star* dated June 16, 2003, written by Faye Casey, and titled *Mystery as 'croc' spotted at pool.*'

The article read as follows:

> 'A Staffordshire community was today trying to unravel a pool monster mystery after reported sightings of a 7ft "crocodile" type creature rising from the deep. Police officers, RSPCA inspectors and an alligator expert from Walsall were called to the pool in Roman View, Churchbridge, Cannock, on Saturday afternoon when reports of the sighting were first made.

'They searched the area and found nothing, coming to the conclusion that the creature must have been a fish or possibly a snapper turtle. But locals are not convinced and youngsters have designed their own "croc on the loose" posters to stick on lamp-posts. One man, who did not wish to named, said he called the emergency services because what he saw in the pool was not a large fish. He and members of his family had being feeding the swans when the creature emerged.

'"We were there looking at the two swans and their baby cygnets," said the man. "And there was a commotion in the water and lots of turbulence. "It was far too big to be caused by a fish. As the creature went past I saw it had a flat head, a 5ft long body, and 2ft tail. It was not smooth and was moving in a snaking action - my initial reaction was it was a crocodile or alligator and so I called the police."

'Linda Charteras, from nearby Cheslyn Hay, was also feeding the swans on Saturday afternoon. "I saw the creature first - a large pool of dirt came up. It looked as though it was after one of the cygnets. I saw its head and long nose and thought there was no way it was a fish," she said.

'Natalie Baker, who lives on nearby Nuthurst Drive, said her children and their friends had been designing the posters. "There has got to be something in it for the police and RSPCA to come out."

'But despite growing local interest in the creature - a group [was] out with their binoculars scanning the water last night - the RSPCA say it is highly unlikely the beast was an alligator or croc. Nick Brundrit, field chief inspector for the RSPCA, said the team kept up observations at the pool for around an hour and a half on Saturday, but there were no obvious signs of an alligator-type creature. He said the sighting was more likely to be a group of basking carp swimming together, or possibly a snapper turtle.'

And so, the scene was now firmly set for the CFZ's adventurous quest to begin in absolute earnest. Once again, over to Jon Downes:

'Following on from the excellent preliminary fieldwork carried out by Mark Martin, the main CFZ expedition finally reached Cannock in the early afternoon of 21st July. After a rendezvous at our digs, the Exeter contingent and Mark Martin drove in convoy to the pond at the end of *Roman View*. No matter how many times one carries out an expedition like this to finally see the location of a series of mystery animal reports for the first time. The pond where the crocodile had been reported was surprisingly wild looking; an oasis of sanity in an increasingly desolate and unattractive West Midlands environment.

'Especially considering that on the far side of the pond from where we set up our temporary base camp, a new section of the M6 was under construction. And with what looked as if it had once been virgin woodland on the hillside opposite had been flattened, in order to build a featureless and rather nasty out of town shopping centre, the ground immediately surrounding the pond looked even more inviting.

'A wide range of butterflies and other flying insects fluttered, hovered, and buzzed their way around the thick vegetation, which was about 800 yards long and 300 yards across and which was fringed by reeds and bull-rushes. A contemplative-looking Heron sneered down at us from a large bush at one end of the pond, and - indeed - most of the weekend there gazing down at us with a particularly supercilious manner. The pond was also home to a pair of swans and their three cygnets that cruised up and down the water like majestic galleons were and totally ignored us for the duration of our stay.

'From CFZ HQ in Exeter came me, Richard Freeman (who had only been back in the country for four days after his first expedition to Sumatra), Graham Inglis, John Fuller, and Nigel Wright (on his first CFZ expedition for some years). We were joined by the aforementioned Mark Martin, Peter Channon (from the *Exeter Strange Phenomena* group), Chris Mullins (from *Beastwatch UK*), Neil Goodwin (from *Mercury Newspapers*), and Wilf Wharton (the CFZ Wiltshire representative who was soon to be immigrating to the Antipodes). Much to my amazement, everybody turned up roughly on time, and we gave three short briefings: one from me, giving a general overview of the events; one from Mark who provided additional background data; and the third from Richard, who cautioned on the do's and don'ts of handling crocodiles.

'I split the available personnel into three field groups. There was the Boat-Team (Mark and Graham); The Away-Team (Richard, Wilf, Chris, Neil and Peter); and the Shore-Team (me, John and Nigel). The initial idea was that the Boat-Team would spend Monday and Tuesday carrying out intensive sonar sweeps of the lake, with the intention of determining the depth any large fish or errant crocodilians. In the meantime, the Shore-Team would scour the shoreline in search of signs of a large beast and also to determine the entry and exit points of the pond.

'Even as John, Graham and Mark struggled to get our trusty dinghy, *The Waterhorse* (named after Loch Ness Monster-hunter Tim Dinsdale's boat), inflated and onto the water, the first set of eyewitnesses arrived. They were a motley gaggle of teenage boys who came up to us; and in thick Brummie accents asked: "whether you're here for the crocodoile, loike?" We replied in the affirmative, and they told us that they had also had an encounter with a scaly creature in Roman View Pond. Richard quickly interviewed them.'

So, with that last point in mind, before we return to the words of the good Mr. Downes, I have related below the entire interview between Richard Freeman and the teenage boys:

RICHARD FREEMAN: 'I gather that you've actually seen this animal and fed it? Could you please tell us exactly what happened?'
LADS: 'We came down at just after the RSPCA had been here. We saw what looked like the animal in the water; and so Elliot went and got some chicken and we lobbed it into the water to feed it. Some of it went too far away. But then we threw one-piece and it landed just next to it and there was a massive splash and we could see both the head and the tail. We actually

thought that we had seen two of the animals in the water but then remembered there is at least one massive pike in here and that the other animal was a fish.'
RICHARD FREEMAN: 'And what did this animal look like?'
LADS: 'It was dark and about five-feet-long including the tail.'
RICHARD FREEMAN: 'Did you see scales or ridges on the tail, or anything like that?'
LADS: 'We didn't see its tail properly but there did seem to be a few spikes.'
RICHARD FREEMAN: 'And have you seen the animal since?'
LADS: 'No. We stopped coming down here after the TV people had been. We have been told to keep away from the pond by some of the local residents.' (Note from Jon Downes: 'There then followed an amusing teenage rant about one of the women whose house overlooks the pond and whom the gang of lads seem to cordially dislike, before Richard managed to bring the conversation back on course.')
RICHARD FREEMAN: 'Do you know anybody else who has claimed to have seen it?'
LADS: 'Yeah, a couple of our friends. One of our friends had been out walking her dog and spotted it. This was the first time that it was seen. Also, a lot of kids from our school have been bunking off at lunchtimes it and coming down here. Some of them say that they have seen it. One day, we came down and there were about fifty kids sitting on the bank.'
RICHARD FREEMAN: 'Do you know whether it has ever been seen on land?'
LADS: 'Not to our knowledge. No.'
RICHARD FREEMAN: 'Does anybody - not necessarily you - have any ideas about where it might have come from? '
LADS: 'We were told that it might have been a pet that got too big and was thrown out.'
RICHARD FREEMAN: 'There seem to be a lot of little streams and pipes which come in and out of the pond. Do you have any idea where they go to?'
LADS: 'Not really.'
RICHARD FREEMAN: 'Prior to this there has not been anything odd reported in this lake before?'
LADS: 'I don't think so. I seem to remember that there was some speculation about something in this pond a few years ago but can't remember the details.'

And with the interview complete, let us once again return to the words of Jon Downes: 'The group of teenagers went about their business, and we went about ours. However, at least at first some of the other local residents were not as friendly. From the moment we arrived the net curtains began to twitch, and soon a procession of local residents walked past us - nonchalantly - to find out what we were doing. Nigel spent much of his time in conversation with these people, explaining the details of our mission and reassuring them that we were perfectly harmless.

> 'There was one slight problem however. Despite having made every effort to contact the owners of the pond (we had even instituted a search with Her Majesty's Land Registry), we had been unable to find them. After we had been at the pond for less than an hour, one irate local who claimed to be a friend of the owner approached us in a combative and pugnacious manner. For a brief few moments it looked to us we were going to be embroiled in an unpleasant scene. However, John Fuller and I managed to calm the situation down, and the man

disappeared, reasonably mollified.

'Finally we managed to get the boat onto the water and the away team was dispatched to the far side of the pond. Then pay-dirt! Nigel, by luck more than by judgment, ran into the lady whose family had been renting the property for 38 years. She could not have been more helpful; and despite the fact that we were trespassing on her property, she granted us permission in writing to carry out whatever investigations we felt were necessary.

'At about 6.15, after a series of false alarms, Mark Martin - in the boat - had a sighting of what appeared to be the 18-inch long dark blackish green head of a large animal. It was not a positive sighting of a crocodile; but it was the best that we had managed to achieve. At the same time, the away team found an area of flattened reeds, which had looked as if a large animal had made itself comfortable; after emerging from the waters of the pond. Unlike other such areas around the shores of the pond, there were no downy feathers from one of the swans; and as the area of flattened vegetation was too big for any known mammal species from the area, it seemed quite possible that this had been the resting place of our mystery crocodilian.

'As soon as we had permission to survey the pond and its surroundings, and we were now no longer conducting a covert operation, we laid a series of navigation lines across two sections of the lake. We took a series of sonar readings to determine the depth of the lake along the lines and found to our surprise the depth of the lake seemed to change by the minute. The next day we found that the lake was fed by a series of sluice-gates from connective channels which crisscross the entire area. We discovered that the bottom of the lake was mostly fairly thick silt, and found that the influx of water from the north end was causing waves in the silt itself, which meant that the depth of the lake fluctuated in some places from between 2.5 and 4.5 feet. Then in the early evening, John Mizzen, one of the original witnesses who had been interviewed by Mark Martin, turned up.'

Yet again, we turn the reins over to Richard Freeman:

RICHARD FREEMAN: 'Basically, can you recount the story from scratch?'
JOHN MIZZEN: 'We were over on the other side of the pond feeding swans, when about five feet from the water's edge my daughter-in-law was looking down this way while I was looking at the lake. She saw the - what ever it was - and said "That's never a fish." It then swam along the water's edge, where I reckon that the water is no more than two-feet deep and it was about five-feet-long and that's including the tail. When it got five or ten feet away from us, it came up and broke the surface. Its head was flat, as was its jaw and its nose, and it was dark greenish black in colour and about eighteen-inches wide. The tail had a scaly appearance, and then it went underneath the water and we just lost contact with it. It had been on the surface for about three or four seconds and in that time it covered about fifteen to twenty feet.'
RICHARD FREEMAN: 'On its head did you notice anything about the eyes?'

The Mystery Animals of Staffordshire

JOHN MIZZEN: 'I didn't see anything of that; not the eyes sticking out of their head or the water or anything. I only saw it from behind and the surrounding parts to its eyes were not visible as far as I could tell.'

Jon Downes once more picks up the story:

'Later that afternoon Richard spoke to a number of other elderly gentlemen who requested anonymity. One of them told us that there had been a series of incidents at a slaughterhouse which was on the shores of one of the other ponds connected to Roman View Pond by a watercourse. Apparently this establishment - which dealt predominantly with the dispatching of elderly and ill horses - supplied meat to local zoos. Some of the meat was hung in a concrete pit in order to prepare it for consumption by zoo animals. Whilst it was hanging something had taken enormous bites out of the carcasses. On another occasion a horse was attacked. Apparently, in the vicinity there is a training-stable, at which horses learn to draw old-fashioned hearses. One of the ways that they trained these animals to walk slowly is to swim them in another of the local ponds, which is connected by a watercourse to Roman View Pond itself. On one occasion whilst one of these horses was swimming, it was attacked by something. When they got it out on to the bank it had a massive bite on one of the back legs. It was eight-to-ten-inches deep and went right down to the bone. The horse was immediately taken to the knackers-yard and shot.

'By this time it was beginning to get quite late in the evening, and so the team then decamped to the local pub, by way of one of the most unpleasant tasting fish suppers that it has been my misfortune to eat. Later in the evening, as it was approaching dusk, we returned to the pond and spent three hours searching the surface of the pond with three one-and-a-half-million candle power spotlights. The Away-Team, with head-torches strapped on, scoured the bank, and out in the middle of the lake Mark and Graham sat patiently in the boat waiting for a scaly monster to surface. Needless to say all these searches were fruitless and at about one in the morning we packed up for the night.

'The next day the CFZ posse was all up and about relatively early. After an excellent breakfast we arrived at the lake soon after 10 o'clock in the morning. Within twenty minutes everybody else had joined us (except for Wilf, who had been forced by the pressure of work to drive down to the south at the end of the previous night's escapades).

'In many ways the second day was a slight disappointment after the adventures of the first. It seemed like that. Although, when you look back, it's now easy to see that we achieved even more. However, at the time it didn't feel like it. Whereas on the first day we had been rushing about, and we had even logged a sighting, much of the second day was spent hanging about, waiting for something to happen.

'The boat party continued their sonar sweeps of the lake, while the shore party continued their explorations of the bank in search of footprints and signs of crocodilians. Sadly, no such signs were found. Indeed, although on the previous day, we had managed to log one pretty good sighting by Mark Martin, today we had none at all. However, this did not mean that the day was a complete waste of time.

'In the original newspaper report, a local lady called Natalie Baker was quoted as saying that her children and their friends had been so excited by the media activity following the initial crocodile sightings that they had spent some time making coloured posters of the animal as part of a school project. Now, Nigel has been working with and for me and nearly seven years now, and over the years I have asked him to do some extraordinary things for me. I have never before said to him "Dude, I want you to find me a little girl who draws pictures of crocodiles." But I did, and - not at all to my surprise; because over the years I have known him I have come to rely on his powers of deduction a great deal - he not only found me the little girl, but managed to persuade her to give me one of the aforesaid posters. Flushed with success after that particular triumph, Nigel and I went off in order to try and solve another mystery, which - we felt - was likely to have a pivotal importance in solving the case of the Cannock crocodile once and for all.

'Richard and I have been members of what I like to call the "UK Animal Mafia" for some years. This is a weird sort of freemasonry that consists of people on the fringes of the pet trade, the zoo trade, and the professional zoology trade. These people - even when it would seem that they have completely opposing agendas - often co-operate to a surprising extent. One of the foremost members of the Zoo Mafia in the Midlands had warned us about the activities of a particularly unscrupulous reptile dealer who was – allegedly, at least - operating in the Cannock area. Nigel and I left the shore party and the boat party doing its own respective things and went undercover.

'It was surprisingly easy to track this fellow down. He had left a trail of debts a mile long; and whenever we went we couldn't find anybody who would say a good word about him. We found the shop where he had once operated a business, which - according to one of our informants - had been closed down on animal welfare grounds. We spoke to his erstwhile landlord and found that when he closed he had left large sums of money owing. We found that he had then set up business under another name in another part of town, but this too had gone the way of all flesh. After two failed businesses, we discovered that the person in question had most recently been sighted working part-time for a pizza delivery company, and selling the remnants of his stock through small ads in the local paper.

'Although we cannot prove it, we were convinced that this discovery had essentially solved the provenance of the Cannock crocodile. It was obvious that somebody had been dumping exotic reptiles in the district. Only a couple of days before we arrived, the *Wolverhampton Express and Star* had carried a story about a large common snapping-turtle which had been captured in a local brook.

'Although the newspaper report claimed that the turtle - named "Lucky" - by the RSPCA inspector who captured him could have been over 20 years old and had "probably lived most of his life in the wild", having inspected the brook in question, and furthermore knowing that when snapping turtles achieve the size of the specimen fished out of this tiny brook in Staffordshire they are very sedentary creatures, who on the whole sit on the bottom of a stream waiting for something to swim into the open mouths, I feel it is far more likely that "Lucky" was dumped into the stream in question within the last few weeks.

"Feeling rather pleased with ourselves, for having completed what we regarded as a rather tidy piece of detective work, we returned to the lake. We discovered that in our absence the CFZ operatives whom we had left behind had discovered some useful data about the age of the lake. Apparently, it had begun life as a pit from which locals dug coal. When the coal petered out, in the mid-1930s, it had begun to fill with water. However, it was a long and slow process, and it wasn't until after the war that the water was deep enough to swim in.

'We also spoke to one of the head-honchos of the local angling society and we discovered that although there were some very big carp in the pond, the largest pike that anyone had managed to catch was only about 9lb in weight. However, according to the local water bailiff there was at least one massive pike weighing in excess of 23lb and probably more than three-and-a-half or four-feet in length.

'The Shore-Team had also managed to identify a number of other small ponds in the area and had found of that most of these were interconnected - either by culverts or by open-water courses. One of the strangest things that we discovered was that somebody had been dumping koi-carp into several of these ponds. As some of you may know, I used to write a column for *Koi Carp* magazine; and so with these very limited credentials Nigel, Richard, John and I paid a visit to a small koi-farm about half-a-mile away. They too had heard the stories about koi-carp - some of them quite sizeable and worth quite a lot of money - being dumped into these local ponds. But there they were completely unable to let us know who had been dumping them and why.

'The next day, we found ourselves in the middle of Cannock Chase, and deep in conversation with the local wildlife-officers who told us that koi had also been turning up in isolated ponds across Cannock Chase, as well. It seems as if there is some kind of strange, Piscine Johnny Appleseed at work, doing his best to stock of the waterways of the West Midlands with these large, ornamental fish.

'Back at the pond we were ready to do a reconstruction of the original sighting by John Mizzen, Linda Charteris and her children. Some time before, we had instituted the practice of performing reconstructions of sightings filmed from two or three different angles; much in the manner of the BBC television programme *Crimewatch*. We have found that using these methods is an invaluable tool in field investigations; and although we had already interviewed both John and Linda in some depth - as had Mark right at the beginning of the investigation - we decided to carry out one of these reconstructions at the pond. We filmed it from three angles; Neil on one side, Mark on the other and Graham filming from the boat. It is always interesting carrying out one of our *Crimewatch* reconstructions and we have never yet done one where we didn't learn something new.

'John Mizzen is probably one of the most professional and accurate eyewitnesses that it has ever been my pleasure and privilege to work with. During our *Crimewatch* reconstruction we discovered that his estimates of the distance that the crocodile had been from the shore and our measured distance differed by only a few inches.

'After the *Crimewatch* reconstructions, we slowly began to break camp. John and Neil lit a barbecue, which had been donated to us by Chris Mullins, and soon the fragrant smell of slowly charring burgers drifted over the evening wind. Someone produced the remains of a bottle of Scotch, and Nigel appeared from *Sainsbury's* with two dozen bottles of beer. The CFZ drank, ate, and watched the sun go down. Neil disappeared back to Liverpool, and the rest of us went down the pub.

'Tomorrow was another day; but, unfortunately we had not caught a crocodile. From the eye-witness descriptions, Richard and I are fairly convinced that we are talking about a spectacled caiman of between three and five-feet in length. Sadly - unless it is very lucky, and somebody manages to fish it out of one of the connecting streams - it is doomed to a slow and ignominious death as soon the first chills herald the advent of the season of mists and mellow fruitfulness. And all because of some stupid selfish Bastard who wanted an exotic pet! *C'est la vie*; unfortunately.'

The story is not quite over, however.

Six years later, in 2009, it seems that someone was yet *again* up to their old, nighttime tricks, and once more anonymously dumping exotic creatures in a certain body of water around the town of Cannock. On this occasion, however, the scene of all the action was a small, three-metre-deep pool that is hidden in a corner of the Brickworks Nature Reserve at Wimblebury - which is only a stone's throw from the heart of the Cannock Chase.

As the *Chase Post* newspaper humorously noted, up until recently 'the only things lurking in the murky waters were six bicycles, a shopping trolley and scaffolding poles.' But all that changed in 2009. Cannock Chase Council officials, concerned about vegetation dying, made a startling discovery, said the *Post*; adding that amongst the usual debris and rubbish, 'there were fish in the water, lots of fish - 20,000, to be precise. Even more baffling, there were not just native species: as well as roach and perch, ornamental varieties such as brown goldfish and koi carp were found.'

The 'Post' expanded further:

> 'Ray Smythe, clerk at Heath Hayes and Wimblebury Parish Council, said: "No-one knows how on earth they got there. We can only think someone released them, but I'd be surprised if anyone knew the pool was there."'

The newspaper noted further that:

> 'Members of Stoke-on-Trent Angling Society have been drafted in to net the mystery fish - and move them to nearby Milking Brook. A spokesman for the club confirmed the operation had been a success. He said: "We estimated that around 20,000 fish were transferred to Milking Brook. This needed three journeys, which, in each case, involved three tanks full of fish. I can confirm very few fatalities occurred during the operation."'

There is very little doubt - as Jon Downes' and Richard Freeman's fine detective work demonstrated back in the summer of 2003 - that someone was still releasing exotic creatures into the pools of the Cannock Chase as late as 2009. But: whether or not this latest development in the strange saga was directly linked to the earlier activity, or if it's an example of yet *another* person adding to the ever-growing body of out-of-place animals that inhabit the waters of Staffordshire, is a matter that still remains very much to be seen and resolved. Whatever the ultimate answer to that question may ultimately prove to be, there is still another matter that needs to be carefully noted: other exotic beasts have been reported inhabiting the dark pools of the area for many a year.

Norman Dodd lived for a good number of years in Glasgow and, in the 1970s, regularly commuted to various parts of central England on business. It was at some point in the hot summer of 1976 that he had a remarkable encounter on the Cannock Chase – with what he says was a giant snake - or possibly even a monstrous eel. Unfortunately, Dodd cannot recall the exact location where the incident occurred, but he can state with certainty that it was a small pool – no more than twenty feet by thirty, he states confidently – that existed at the time, and that he believes was 'not far from Slittingmill'.

Dodd states that he had parked his car, a Ford Cortina, on the makeshift car-park that was adjacent to the pool and was happily munching on his sandwiches and reading a newspaper.

> 'It was a bloody stifling day – remember that summer, how hot it was? I remember swigging something to drink and having a bite when there was something moving right on the bank [of the pool].'

Dodd adds that he was startled to see a creature that he estimated to be around fifteen feet long slowly surface from the water and that then proceeded to 'bask' on the banks of the pool. 'It sort of wriggled,' said Dodd, 'like its whole body seemed to sort of shake or wobble as it moved.'

Dodd further reveals that the animal had a serpent-like head and oily-coloured skin. Its body was 'thick' and it seemed to be wholly unconcerned by his presence. 'I know it saw me – or saw the car, definitely – because it looked right in this direction and then just went back to what it was up to: laying there.'

But, what was most puzzling to Dodd was that the animal seemed to have 'flippers near the front – or little feet or something.' Dodd concedes that the animal *may* have had similar 'flippers' or 'feet' at its rear too, but he explains that he cannot be wholly certain; since the 'back-end never came right out of the water; like as if it was trying to keep itself cool from being part [sic] in the water.'

Dodd watched astonished – and not a little bit concerned, too – for at least twenty minutes, after which time the animal slid, in a relaxed fashion, back into the waters of the pond. 'I wondered how a small pond like that might feed an animal that big for food [sic]. But what about the feet or flippers? Does that mean it might have been able to go from pool to pool for fish and things?'

But for those who might consider the possibility of huge serpents or giant eels on the loose in the Midlands to be just too extreme, take careful note of the following, published in the *Birmingham Evening Mail* newspaper in January 2003. It may very well cause you to drastically re-evaluate your opinion:

> 'Walkers, joggers and cyclists have been pounding a towpath in Edgbaston unaware of a near 15ft. Burmese Python lurking in the depths just feet away below the water. The oversized serpent, capable of killing a child, has been fished out of the Rotten Park Road canal in Edgbaston by the RSPCA yesterday. It was spotted by a terrified passer-by. It was not clear how long the reptile had been living among the old tyres and shopping trolleys, but experts said it hadn't been dead too long.
>
> 'RSPCA Inspector Rob Hartley, based at the rescue centre in Barnes Hill, said: "It's like something out of a horror movie. This thing is massive; we've never seen one this big before. It's a monster. We've measured it at fourteen and a half feet and up to fourteen inches wide. It probably weighs at least eleven stone. It could kill a child by wrapping itself around it and suffocating it. We don't know whether it simply got too big for someone to look after and they let it go free or it escaped."'

The newspaper added:

> 'Inspector Hartley said that the snake's size indicated it had been well kept by whomever, even though it would have devoured around twenty-four dead mice or day-old chicks a day. He said the Burmese Pythons were relatively common pets, but usually only reached about six feet in captivity.'

And, while we're on the subject of monsters of the deep, I cannot resist mentioning the following story from the *Chase Post* that surfaced in 2006 and that is *kind* of related to the subject-matter of this chapter; albeit in a strange and roundabout fashion! Yes, according to the newspaper, nothing less than Nessie had come to Cannock! But this was not the famous long-necked lake-monster with which Scotland's tourist industry is so enamoured and obsessed.

It was in June 2006 that Cannock's High Green Court received a new resident – a South American redtail hawk named Nessie. Brought in to try and quell the ever-growing pigeon population of the area, Nessie began making regular 'patrols' of the Court's grounds in what ultimately became a highly successful attempt to see off any unwarranted visits from the local pigeon population – which had been blamed for eroding limestone from the town's War Memorial, among other many and varied crimes of an allegedly heinous and national security-based nature.

Not everyone was happy with the situation, however. One Court-worker fumed to the media:

> 'I'm absolutely disgusted that this goes on in broad daylight where children can witness it. I like seeing pigeons when I look out the window. What's wrong with people today? Why can't the pigeons live? Why does everything have to be so sanitised? It has made me feel really sick.'

This Nessie, it seems, was just as controversial as its Scottish namesake – but for distinctly different reasons...!

Chapter XII
Monsters of the Full Moon

'It just looked like a huge dog. But when I slammed the door of my car it reared up on its back legs...'

Even a man who is pure of heart and says his prayers by night, may become a wolf when the wolf bane blooms and the autumn moon is bright,' was the message immortalised in the classic 1941 film from Universal Studios, *The Wolf Man*. And it is a message that many have taken, and continue to take, extremely seriously – and especially so within the county of Staffordshire.

For those Staffordshire folk who believe in the existence of literal werewolves, the image of the hairy shape-shifting beast that is part-human and part-wolf, and that embarks upon a marauding killing spree at the sight of a full moon, is no joke whatsoever. But if such creatures really do exist, and they are indeed prowling around the county by the eerie light of a full moon, are they really true werewolves of the type that have been so successfully portrayed on-screen time and again by Hollywood movie-moguls? Could they perhaps be deranged souls, afflicted by a variety of mental illnesses and delusions? Or might they have distinctly paranormal origins? Paradoxically, the answer to all three of those questions might very well be: 'Yes.'

Before we get to the specifics of the mystery of the Staffordshire werewolves, a bit of background data on the overall phenomenon is most definitely required.

Clinical-Lycanthropy is a rare psychiatric condition which is primarily typified by a delusion that the afflicted person has the ability to morph into the form of a wild animal – and very often that of a berserk, killer-wolf. Interestingly, a 1999 paper titled *Lycanthropy: New Evidence of its Origin* by H.F.Moselhy, demonstrated that two people diagnosed with Clinical-Lycanthropy displayed evidence of unusual activity in the parts of the brain known to be involved in representing how we perceive body-shape and image. In other words, Clinical-Lycanthropes might very well really believe their bodies are wildly mutating when they are overwhelmed by their delusions.

Of course, this does not fully explain why so many such people believe they are changing into one specific animal – namely a wolf – rather than just experiencing random changes in, say, their arms, legs, hands, feet or head. But, nevertheless, it is without doubt a significant part of the puzzle.

And there is another aspect to this affair that may go some way towards explaining the inner-workings of the strange mind of the Clinical-Lycanthrope.

Linda Godfrey, a leading authority on werewolves in the United States, says:

> 'One other medical explanation that turns up frequently in relation to lycanthropy is the ergot equation. A fungus that affects rye, ergot is now widely regarded as a possible cause of the bestial madness. According to this theory, it was not demonic influence but the ingestion of Claviceps purpurea (which contains a compound similar to LSD), which led to the demented behaviour. All it took was a cold winter in a particularly wet or low-lying area, and entire fields would be infected with the ergot fungus.'

She continues:

> 'Symptoms, confirmed by an outbreak as recent as the 1950s in France, include delusions of turning into hairy monsters, night terrors, a sense of alienation from one's own body, frantic motion and convulsions, paranoia, and even death.'

Beyond any shadow of doubt at all, one of the most notorious serial-killers of all time was Peter Stumpp, a German farmer who became infamously known as the Werewolf of Bedburg. Born in the village of Epprath, Cologne, Stumpp was a wealthy, respected, and influential farmer in the local community. But he was also hiding a very dark and diabolical secret - one that surfaced graphically and sensationally in 1589, when he was brought to trial for the crimes of murder and cannibalism.

Having been subjected to the extreme torture of the rack, Stumpp confessed to countless horrific acts, including feasting on the flesh of sheep, lambs and goats, and even that of men, women and children, too. Indeed, Stumpp further revealed that he had killed and devoured no less than fourteen children, two pregnant women and their fetuses, and even his own son's brain. Stumpp, however, had an extraordinary excuse to explain his notoriously-evil actions.

Stumpp maintained that since the age of twelve, he had deeply engaged in black-magic, and on one occasion had actually succeeded in summoning up the Devil, who delighted in provided him with a 'magical belt' that gave him the ability to morph into 'the likeness of a greedy, devouring wolf, strong and mighty, with eyes great and large, which in the night sparkled like fire, a mouth great and wide, with most sharp and cruel teeth, a huge body, and mighty paws.'

The court, needless to say, was far from being impressed at all, and Stumpp was quickly put to death in a brutal fashion: flesh was torn from his body, his arms and legs were broken, and, finally, he was beheaded. The Werewolf of Bedburg was no more. Stumpp was not alone, however.

Equally as horrific as the actions of Stumpp were those of a now-unknown man who, in the final years of the 16th Century, became known as the Werewolf of Chalons. A Paris, France-based tailor who killed, dismembered, and ate the flesh of numerous children he had lured into his shop the man was brought to trial for his crimes on December 14, 1598. Notably, during the trial, it was claimed that on occasion the man also roamed nearby woods in the form of a huge, predatory wolf, where he further sought innocent souls to slaughter and consume.

As was the case with Stumpp, the Werewolf of Chalons was sentenced to death, and was burned at the stake.

While the idea that mental illness, possibly accompanied by the ingestion of ergot, may account for *some* of the legends of werewolf activity is a highly plausible and very likely one, not everyone is quite so sure that it answers *all* the questions. Indeed, seemingly credible witness testimony suggests that the British Isles might very well be home to werewolves of a distinctly flesh-and-blood variety.

For example, early one morning in the winter of 1952, says now 74-year-old Margaret Shelley, she saw with her own eyes 'a very big, hairy man with a wolf's head', roaming near the shores of Loch Morar – a Scottish loch that, notably, also has a longstanding and famous lake-monster tradition attached to it. According to Shelley, at the time, she was there with her fiancé, as the pair was visiting friends in the area, and had decided to take a trip to the loch to pass a few hours. Today, Shelley wishes she had never set eyes on the cursed loch.

Shelley further adds that when she first saw the beast, it was at the edge of the loch, seemingly 'lapping-up the water.' She estimates that at the time, she was perhaps no more than two hundred feet from it. As Shelley moved closer, 'to get a better look', however, the creature reared up to a standing position, swung its head around and stared intently her direction.

'It gave me a big fright,' she says, with understandable justification, adding: 'It was a horrible look; evil, and with big pointed ears. And it was so tall: about seven-feet.'

For no more than a few seconds both woman and werewolf locked eyes on each other, before something truly startling happened, as Shelley explains: 'It was standing upright; but when it went to run away it went down on all-fours like a dog and raced away.'

Shelley, thankfully, never saw the animal again. But this was not the only occasion upon which Scotland has allegedly had a real-life werewolf in its midst: a similar event occurred in 1967 – again, interestingly enough, during the winter months.

In this case, the location was the town of Oban, and the witness was a postman, driving to work in the early hours of the morning. As he headed along the moonlit road in question, he was shocked to the core to see coming in the opposite direction, a tall, man-like figure with wolf-like features. Not only that: it was hurtling along the road at a fantastic rate. 'It was there one second and it had raced by in the next,' says the witness.

The Mystery Animals of Staffordshire

Moving from Scotland to England, there is the story of the Shirley family of Kent. According to Pat Shirley, while picnicking in an area of woodland on the east coast of England in the late 1940s, her grandmother had seen a huge animal that looked like something straight out of *An American Werewolf in London*. In this incident, the beast had what was described as 'flaming red hair all over it', and possessed a pair of huge and powerful jaws. Again, it was only seen for a moment or two before vanishing into the trees.

And, of course, there is the famous story of Northumberland's Hexham Heads – a true tale of supernatural lycanthropy from 1972 that is told in my 2008 book, *There's something in the Woods* – and the many and varied tales of werewolves on the loose in Devonshire that Jon Downes describes in his Gonzo-travelogue *Monster Hunter*. And, with the scene now set, let us turn to the issue of the moon-beasts of Staffordshire.

Midway through 2007, reports began to surface from the seemingly monster-infested Cannock Chase to the effect that nothing less than a blood-thirsty werewolf was on the loose, terrifying the village folk of the area, and ensuring that the Chase became a no-go area after the sun had set and overwhelming darkness had firmly set in.

On April 26, 2007, the *Stafford Post* newspaper sensationally related the following:

> 'A rash of sightings of a "werewolf"-type creature prowling around the outskirts of Stafford has prompted a respected Midlands paranormal group to investigate. *West Midlands Ghost Club* says they have been contacted by a number of shocked residents who saw what they claimed to be a "hairy wolf-type creature" walking on its hind legs around the German War Cemetery, just off Camp Road, in between Stafford and Cannock. Several of them claim the creature sprang up on its hind legs and ran into the nearby bushes when it was spotted.'

The 'Post' added:

> "Nick Duffy, of *West Midlands Ghost Club*, said the stories of werewolf sightings in Chase area were something that he had encountered before. He said: "The first person to contact us was a postman, who told us he had seen what he thought was a werewolf on the German War Cemetery site. He said he was over there on a motorbike and saw what he believed was a large dog. When he got closer, the creature got on his hind legs and ran away. I've spoken to many witnesses and I know when they are putting it on. But what struck me as strange about this was the way he told it. I'm in no doubt that he was telling the truth."'

The beast in question was also reportedly witnessed by a scout-leader who was strolling around the woodland earlier that same month. The witness, who the *Stafford Post* stated flatly refused to go on the record with his name preserved for posterity, said he saw what at first he thought was a huge dog lurking in nearby bushes. However, as he looked closer, the man realised this was no normal dog – at all. He explained thus: 'It just looked like a huge dog. But when I slammed the door of my car it reared up on its back legs and ran into the trees. It must have been about six to seven feet tall. I know it sounds absolutely mad, but I know what I saw.'

The Mystery Animals of Staffordshire

Were genuine flesh-and-blood werewolves really roaming around the darkened woods that surround the German Cemetery? Many people were positive they were, and the local media was not at all afraid to continue to immerse itself deeply in the controversy. As evidence of this, consider the following from the *Chase Post*:

> 'A tribe of subterranean creatures who surface on Cannock Chase to hunt for food could be behind a rash of 'werewolf' and Bigfoot sightings near Stafford. And the mysterious beings could also be responsible for a string of pet disappearances, it has been claimed. Theories behind the sightings range from a crazed tramp to aliens. But now another paranormal expert has put forward the theory the sub-human beast is not a werewolf at all - but a Stone Age throwback. The investigator, who wishes to remain anonymous, told us: "Strange sightings in this area have been made over many years by civilians, military, police, ex-police and scout leaders on patrol. Some incidents have been reported and logged but others not: some people don't want to be classed as mad."'

The 'Post' additionally reported:

> 'The strangest rumour has come from a senior local resident who believes the mysterious intruders to be subterranean. He told us: "The creatures have made their way to the surface via old earthworks to hunt, for example, local deer. It's a fact that there has been significant mining activity under Cannock Chase for centuries. And it's a fact there is a high rate of domestic pet disappearance in the area - especially dogs off the lead. Just ask anyone who walks their dog near the German War Cemetery.'

Of course, with fantastic tales of werewolves, subterranean monsters, ancient earthworks and disappearing pets being highlighted at every available moment, the furore rumbled on at an ever-escalating pace, as the *Chase Post* continued to note with undoubted deep relish and satisfaction:

> 'A team of investigators from an organisation called *Beastwatch* could be setting up camp on Cannock Chase to search for creatures "real or surreal". Following a catalogue of reports in recent months about all kinds of bizarre sightings on the Chase, the Leicestershire-based team has offered to investigate the area. They hope to establish once and for all if anything unusual - mammal or mystical - is stalking the area. Recent reports have included a panther, caveman, werewolf, UFOs and even a floating ghost in the cemetery. But in order to carry out the research they need your help.

> '*Beastwatch*, a voluntary organisation, relies on public donations to pay expenses. And as the investigators hope to set up camp for several days and nights they will need travel and subsistence expenses. The rangers have also informed the team they will be charged for staying on the Chase. *Beastwatch* co-ordinator Chris Mullins told *The Post*: "We would be looking for footprints and laying sand traps. We put bait in the middle of the sand to see what comes. We will also look for underground entrances at various parts of the Chase. If it looks

as if we are onto something we may be there longer."

'Chris, aged 55, described himself as "a person with an open mind who walks with his feet firmly on the ground." He will be bringing infra red cameras and sound equipment.

Chris's team has already visited Sherwood Forest looking for "a Bigfoot-type, eight foot hairy man-beast with glowing eyes" said to live in the many underground caverns. They didn't find him. *Beastwatch* has already visited our patch to investigate claims a crocodile is living in waters in Bridgetown. That investigation did not find the croc - but did not dismiss its presence either. Chris has been involved with looking for everything from wild boar to tarantulas and the more mystical beings for six years. If you would like to sponsor the vigil, or help in any way, contact Chris. But any volunteers will be interviewed first. "The last thing I want is someone who runs screaming out of the woods," Chris added.'

The 'Post' also contacted me for a comment-or-several on the monstrously-odd affair. The 'Post' stated, after I was interviewed by 'Post' writer Mike Bradley:

'Internationally-renowned *X-File* investigator and author Nick Redfern has shed new light on the mystery wolf-like creature said to be roaming the ancient woodlands of Cannock Chase. Speaking exclusively to *The Post* from his Dallas home, Walsall-born Nick said he has been busily amassing credible evidence on paranormal experiences across the Chase.

'And while his research has found that it is "highly unlikely" to be a werewolf or "caveman" lurking in the undergrowth, he believes the cause of people's sightings is of paranormal origin: "I have investigated reports of wild boar, wallabies, big cats and even Bigfoot-like beasts in the area. But the werewolves are definitely the strangest. Back in the eighties, reports surfaced of something that became known as the Ghost Dog of Brereton which was a strange, large dog-like animal. And last summer there were local stories of a wild wolf seen near the M6; so I'm convinced people are seeing something."'

The 'Post' continued to quote me, as follows:

'Nick also takes seriously the idea that a real-life wolf may be at large on the Chase: "We know that people who have had big cats as exotic pets have released them into the wild when they got too big to handle. Perhaps something similar has happened here with a wolf."'

I added that:

'If people *are* seeing some sort of werewolf-like creature, then it's likely to be paranormal in origin rather than physical. I think it is highly unlikely that an evolved, physical creature - a werewolf or caveman - could hide out in caverns beneath the Chase and remain undetected. But people are definitely seeing something out of the ordinary, and it is a fascinating story.'

The *Chase Post* was not done with the Staffordshire lycanthropes yet, however:

> 'Derek Crawley is chairman of *Staffordshire Mammal Society*. He told us wolves, hunted to extinction in England more than 500 years ago, could make a home on the Chase. But he doubts a pack roaming there today could keep such a mysterious and low profile. "There are enough prey items to sustain a population of large predators there," he said. "As for wolves: theoretically, yes but in reality, no. They are a pack animal so people would see a lot of them. They would confront dogs and other animals. So if there was a population, they would very soon be known about. What it could be is huskies. People do train them over the Chase and they can look quite wolf-like." He added: "There are plans to reintroduce wolves but that's in areas like the Scottish Highlands where there are very few people. It wouldn't happen on Cannock Chase."'

And there was another, very strange, matter, too: one of those buried in the German Cemetery was a man named Maximilian von Herff. Born on April 17, 1893, von Herff served as an officer with Nazi Germany's Reichswehr in the First World War, attained the rank of colonel and went on to win the Iron Cross. During the Second World War, von Herff served in North Africa as commander of the Kampfgruppe and went on to join Hitler's notorious and much-feared SS. He died in 1945 in Britain's Conishead Priory Military Hospital, having previously been held at the nearby Grizedale Prisoner of War camp.

It turns out that Nazi Germany had established a clandestine resistance force which would undertake guerrilla-style attacks against the Allies in the event that the Nazi regime came to an end. The group was approximately 5,000 in number and was comprised of members of the SS and the Hitler Youth. Its name was the Werewolves. So, in a very roundabout way, there really *were* werewolves at the German Cemetery – certainly Nazi ones, and very possibly supernatural ones, too.

And the saga is still not quite at its end. In the wake of the publicity given to the tales of the Cannock Chase werewolf, I received the following report from a man named Wes: 'I encountered a werewolf in England in 1970. I was twenty years old when I was stationed at RAF [Royal Air Force] Alconbury. I was in a secure weapons storage area when I encountered it. It seemed shocked and surprised to been caught off guard and I froze in total fright. I was armed with a .38 and never once considered using it. There was no aggression on its part. I could not comprehend what I was seeing. It is not human. It has a flat snout and large eyes. Its height is approximately five feet and [its] weight [is] approximately 200 pounds.

> 'It is very muscular and thin. It wore no clothing and was only moderately hairy. It ran away on its hind legs and scurried over a chain link fence and ran deep into the dense wooded area adjacent to the base. I was extremely frightened but the fear developed into a total commitment of trying to contact it again. I was obsessed with it. I was able to see it again a few weeks later at a distance in the wooded area. I watched it for about thirty seconds slowly moving through the

woods and I will never forget my good fortune to encounter it and to know this "creature" truly does lives among us.'

Wes just may be right. And it is not just in the depths of the Cannock Chase that the Staffordshire werewolves roam, however.

Alrewas is a large village and civil parish situated approximately five miles north of the city of Lichfield and has a population of approximately 3,000. It lies adjacent to the A38 road, which follows the line of Ryknild Street, an old Roman road; and according to the *English Place-Name Society* the village's old name translates as: 'Alluvial land growing with alder-trees'.

Certainly, Alrewas is steeped in the world of the past: it is home to an All Saints Anglican church, which can be found just off Church Lane, and which dates from the 12th century. Some of the original Norman work on the church can still be seen; however, a great deal of Gothic enlargement is also in evidence, while the church font dates from the 15th Century and the pulpit from the 17th Century.

And it was within the confines of this pleasant little village - that one might easily see playing centre-stage in a *Miss Marple* mystery or in an episode of *The Midsomer Murders*, *Lovejoy* or *Jonathan Creek* - that, in the 1950s, a fully-fledged werewolf was said to lurk.

The story comes from a man named Sid Lavender, who, in 1953, was working in the nearby locale of Barton-under-Needwood. At the time, Lavender was twenty-two, and had recently completed a stint of National Service in the Royal Air Force. Lavender was new to the area, and at the time of the incident in question, had only made one friend: another young man who had also then recently completed his national service and who lived in Alrewas.

Unfortunately, beyond identifying his friend as being named Terry, Lavender is highly reluctant to reveal his full name, 'on account of that I lost touch with him years ago and don't want to put words in his mouth'. Nevertheless, Lavender *is* willing to relate the basic details of the strange story told to him by Terry one winter's night in a local tavern that has long-stood in Alrewas: the 17th Century *Crown Inn*.

So the tale goes, Terry was cycling to work on a freezing December morning around 7.00 a.m. when, from a distance of a couple of hundred feet he saw on the fringes of Alrewas what he initially thought was a 'tall man in a big, long coat, stood dead still' at the side of the road 'where there were trees all around'.

As Terry cycled past, however, his curiosity turned to overwhelming fear: the 'tall man' was actually nothing of the sort. Rather, the being was a 'furry, black animal stood upright with a big, long snout – like a dog, and with dog-ears: pointed', says Lavender.

Ominously, as Terry flew past the beast as quickly as his legs could turn the wheels of his trusty old bicycle, he heard it offer a low, guttural growl 'as if it was saying to him: "just keep pedaling and keep away". Perhaps wisely, that is precisely what Terry did, only looking back several times to see that the creature had not moved an inch, aside from turning its head to keep a close and watchful eye on him until he was lost from sight.

That night, which was a Friday night, Sid and Terry met in the *Crown*, and Terry revealed the shocking details of the strange story to Sid – who, at first at least, considered it 'a good chuckle and nothing else, really'. It was only after Terry 'kept on and on' that Sid realised his friend was being deadly serious with him.

Of course, being young lads, and looking for a bit of adventure on a Friday night, they decided to stake-out the area for a couple of hours - 'despite that it was bloody freezing and we could have stayed longer [in the pub] and had a few more pints' – in the hope that the animal might put in a reappearance. It did not. And tactful questions posed to people in Alrewas did not reveal any more data, either. It seems that whatever the beast was, Terry had been the sole witness to its brief and unearthly presence.

A study of the available evidence – or, rather, the unfortunate lack of it – does not provide any further data on sightings of werewolves in and around Alrewas – either before or after 1953; however, if you ever decide to pay a visit to the picturesque little village, take care as you walk its winding lanes, and particularly so if the sun has set and the moon is full.

One final story that may (and I stress the word *may*) be relevant to the controversy surrounding the werewolves of Staffordshire comes from the Jacoby family, who lived in Leek in the 1970s, and who 'for a couple of months' in early 1972 were 'constantly bothered by a bloody weird howling every night, outside'. Mr. Jacoby says: 'It wasn't like your average dog: it was more like a wolf like you'd see on the telly in one of them nature programmes. It was the sort of howl you didn't forget and don't get from a dog. But I can't think of what else it was. It *did* sound like a wolf, but where would it have come from and where did it go?'

The beast of Leek, it seems, was as mysteriously elusive and transitory as the rest of Staffordshire's werewolves; unless you know better, of course...

Glen Vaudrey
Me again, so you've read about the werewolves that Nick found, now it's my turn to introduce you to some other staples of the Hammer House of Horror world.

Well it does seem that Staffordshire has more than its fair share of werewolf sightings. The Fringe Weird Report noted that there had been 20 werewolf reports from Cannock Chase within the last 25 years, that's certainly a lot of them considering there is only one other sighting recorded elsewhere in the United Kingdom during that time. Of course that's not say that there hasn't been a long tradition of werewolf sightings in the UK, I myself looked into an account of one in my book *The Mystery Animals of The Western Isles* if you don't want to know what my findings were look away now.

(it's almost certainly a hoax)

you can look back now.

The Mystery Animals of Staffordshire

The story as reported from the Isle of Lewis was that the remains of a werewolf were recovered from a shallow grave; the Outer Hebrides was not the only place in the country to have a similar story with tales from both Lincolnshire and Merionethshire. While these three tales reported a body being found no actual body has so far appeared to back up the stories, however in Staffordshire a body does turn up, but the question has to be was the body really that of a werewolf. As you may have guessed it's the next thing that I am going to have a look at.

It was back, well, within living memory just, in April 1975 at 6 o'clock one morning that the body of Andrew Prinold was found dead at the crossroads by a passing postman (didn't those postmen get out early in the 1970s). So how does the body of this unfortunate man bring us to that of a werewolf? Well it is the victim himself that is the connection for he considered that he was turning into a werewolf.

By the time of his inquest details of the case had been published in various newspapers and I have gathered together the majority of them to provide the following account.

Andrew Prinold was a 17-year-old apprentice joiner who had started to attend séances following the death of his father in the hope of being able to contact him. In the sensationalist way that newspapers are known for, it was stated that this interest in séances was prompted by Prinold having seen the film *The Exorcist*; while now it might seem a little tame by the standards of today's horror films in the mid 70s it was about as bad as it could get, a sure route to damnation. The séances themselves, as the inquest was told, consisted of five lads putting a glass on a table with names and numbers around it and then starting to talk to it. It only took a few minutes before Andrew was starting to shake and was sweating and moaning, that seemed to bring to an end the first go at a séance, but a mere ten minutes later they had another go, this time however Andrew took no part in it. Not that that did any good as once again he started to shake. His workmate Stephen Williams told how Andrew suddenly announced that the Devil was inside him. It seems that things were about to get out of control, that was until Willams punched Andrew on the jaw. While it might seem a little unfriendly it worked in *Airplane* to stop panic and it also seemed to work in this case as Andrew was reported to have forgotten it all, as well as being momentarily unable to see, perhaps it was a very hard punch. Eventually Andrew regained his sight and everyone set off on their motorbikes. Like in all good horror films it appeared all was well, but as the coming days would reveal that wasn't the case because a few days before he dies Andrew 'phones Williams and tells him that his 'face and hands were changing colour and were turning into those of a werewolf'. Williams recalled that Andrew would at times go quiet and then start growling, Andrew also told Williams that he had a knife and that he was going to kill himself. The inquest determined that Andrew Prinold did indeed kill himself. Whether the young man was really turning into a werewolf or if it was all just in his mind we will never know but I do find it interesting that he was found dead at a crossroads, a traditional place for the burial of suicides and those who it was thought would arise from the grave.

Hopefully you've now had your fill of the werewolves in Staffordshire so it must be time to move on to that other creature of dark romance novels: the vampire. As luck would have it, also in the 1970s, there was evidence to suggest that there was one person at least who believed that there was a vampire in The Potteries, not to be mistaken for a vampire in the pottery which is an altogether different story.

Sadly, in this report we don't have a sighting of a vampire to work with but rather we have the remains of the victim, or rather a person who appeared concerned not to become a vampire's victim. The event took place in late 1972 in Stoke-on-Trent in one of a group of Victorian mansions known as The Villas, a once rather impressive set of 24 dwellings designed in the Italianate style by the local architect Charles Lynam. When originally built theses houses even had rooms for the servants to live in (those were the days). However by the 1940s these large buildings had started to be sub-divided. In April 1972 the estate was designated as the first conservation area in Stoke, however it's the events of November of that year that we are interested in.

It was 6pm on Friday 24th November, a dark and dingy night, when Police Constable John Pye was called to number 3 The Villas. At the time number 3 was a boarding house run by Mrs Rodziewicz, a Polish lady known locally as Mrs Rod (a name far easier to pronounce). Its many rooms were occupied by folk from Eastern Europe, these were mainly Polish men who had come to the United Kingdom in the aftermath of the Second World War having lost everything including their country to the Soviets. Conditions in the rooms could be a little cramped and it wasn't unknown for three men to share a room.

Constable Pye had been called to the property as the landlady Mrs Rod had not seen one of her lodgers leave his room for a number of days, and furthermore he was not responding to her knocking upon his door. The man in question was Demetrious Mykicura and it appears he was not a Pole but a Ukrainian. Why does it appear so, you may ask. Well, from his un-Polish Ukrainian name, his regional vampire belief, and from my personal correspondence with John Pye.

Mr Mykicura was thought to be around 68-years-old and a very lonely and solitary man with few friends; he had had come to the country after the war and had found work in the pottery industry.

PC Pye was led to Mr Mykicura's room down a long dark corridor lit only by a bulb that he would later equate to being of the lowest wattage possible, finally he arrived at the door of the room. A light tapping upon the door did not get a response, so he announced,

  'It's the police. Sir are you okay?'

Not getting a response he tried again louder this time, once again nothing.

Turning to Mrs Rod he informed her that he would have to break the door down, putting his shoulder against it it soon swung open. At once a musty slightly sickly smell escaped from the room. The room itself was in darkness, at first PC Pye reached for the light switch but before he could find it Mrs Rod told him that there was no light bulb in the room, Mr Mykicura was frightened of electric fearing that it could cause an explosion.

Handily PC Pye of course had his torch with him and he was able to illuminate the room in front of him, and he discovered there wasn't much of a room to see. It was barely eight feet by six and contained a single bed and a tiny wardrobe. As if to add to its charm the wallpaper was hanging from the walls, five star accommodation it was not.

It was the sad-looking bed that drew attention as there was clearly something under the sheets other than just a stack of pillows and loose bedding. Taking a deep breath the PC pulled back the sheets to reveal Mr Mykicura, he lay wide-eyed with congealed froth around his mouth, his arms up by his head, fists clenched, it was clear that he had been dead for a number of days.

Pulling back the covers a little bit more Pye could see by the power of his torch that there were several small cloth bags in the bed alongside the corpse; each bag was tied at the neck with string and all seemed to contain powder. One of the bags was torn and from it trickled out what looked like salt. The bags weren't the only odd things to be found in the room, scattered all around in the bed, on the bed and on the floor were dozens of silvery paper-like fragments that reflected back the light from the torch, on picking up one of these slivers PC Pye discovered that they were pieces of garlic skin. A further look around confirmed that there was a number of garlic cloves about as well.

Faced by this rather puzzling set of garlic and bags PC Pye asked Mrs Rod what was the reason, she replied in her heavily accented English,

> 'It woz for ze vampire, Dimitri believed in ze vampire, zeese things would protect him.'

Now it would be fair to say PC Pye wasn't expecting that for an answer.

While there was no sign of foul play that didn't mean that was the end of it, there was still work to be done recording the scene, not only photographs to be taken but there also needed to be a detailed plan made of the room. It was while doing this that another discovery was made, this time out through the small window. Outside the window there was a small flat area that joined the two parts of the pitched roof, around a foot below the window ledge was an upturned plastic washing bowl, and upon lifting this up there was another surprise to be found: a piece of human excrement. If that wasn't strange enough there was something sticking out from it, by the beam of the torch light it could be seen that these were pieces of garlic stuck in it. Yes, stuck in after the event if you like, certainly not digested.

After talking to Mrs Rod and the numerous Polish residents it was clear that the late Mykicura was a man with, as it turned out, an unhealthy interest in vampires. Eventually the ambulance came to take the body away, and that was that for the night.

As it turned out this case was that unusual that PC Pye wanted to know a little bit more about the subject of the deceased's fears. He went to the library and got his hands on *The Natural History of the Vampire* by Anthony Masters.

The following Tuesday he received a call from the coroner's office who told him that Mr Mykicura had died of asphyxiation. Not strangled as you may have thought but rather he had choked to death. The pathologist reporting that he had died as a result of having a pickled onion in his throat.

Recalling what he had read in the vampire book Pye asked if the object in the throat could have been a bulb of garlic rather than a pickled onion. On a second look it was indeed confirmed to be a bulb of garlic.

Thanks to reading the vampire book PC Pye had learnt a few things that helped put the strange evidence into context. For a start the reason he knew it would be garlic and not an onion that had been the object the victim had choked on was that traditionally garlic had been used to ward off the attentions of vampires. For example, a bulb of garlic in the mouth would provide a barrier that would stop a vampire's spirit entering the body. Of course as we have seen it does have its risks, while choking to death is one of them a lack of friends could be another.

As for the rather charming offering under the washing bowl, well while it might have been seen as a surprisingly effective device to dissuade burglars from breaking in via the window it was actually yet another way to hinder an undead bloodsucker. For reasons best known to vampires it seems some like to devour human excrement (each to their own). Of course a cunning fellow may place pieces of garlic within the not so tasty treat and the vampire would get a shocking surprise tucking into its meal; for a more palatable idea of the fact think of a donut with razor blades hidden inside it. It's not nice however you look at it.

As for the bags of salt that were found, well this was to exploit another of the vampire's weaknesses for it is said that a vampire is compelled to count every grain of salt before moving on to the victim. With enough bags of salt you could keep a vampire occupied long enough to escape from it, the idea was that a big enough pile would keep any wandering revenant busy while you had time to calmly walk away.

The coroner stated that it was not unusual for people to 'bolt their food and die' (that's a worry for speed-eater's everywhere), and at the conclusion he summed up that it was 'this poor man's obsession with vampires that led ultimately to his death, so perhaps they got him in the end.' And with that a verdict of accidental death was recorded.

So we never did find out if Mr Mykicura's fears were founded, but it should be noted that some two years previously there had been talk of a vampire attack on people just forty miles away in Winsford, Cheshire, not that far as the bat flies. You will be able to read all about that in my next book *The Mystery Animals of Cheshire*.

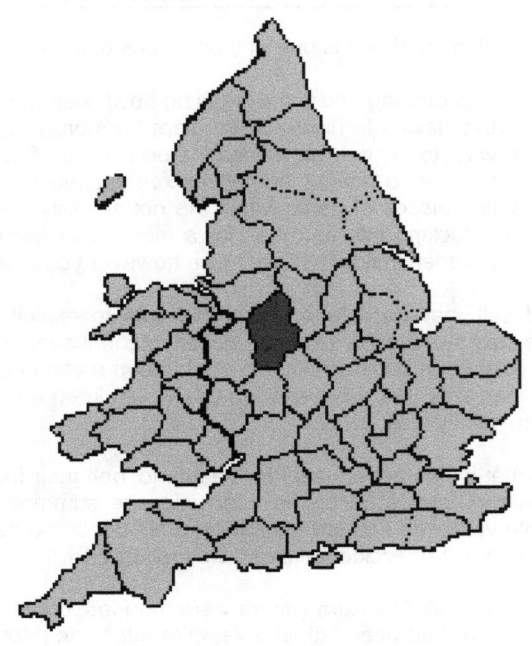

Chapter XIII
Wallaby Wonders
'...I know where they are, but can't tell you...'

While the very idea that living, breathing wallabies are excitedly bounding and bouncing around the wilds of Staffordshire might, to some readers of this book seem manifestly absurd in the extreme, this is a mystery that is actually not so mysterious as it initially appears to be, after all. And, it's a saga that is absolutely true and utterly verifiable, too.

A wallaby is any of approximately thirty species of macropod (*Family Macropodidae*): an informal designation usually cited for any macropod that is smaller than a kangaroo or a wallaroo and that has not been given another name. Very small forest-dwelling wallabies are known as pademelons and dorcopsises, and the name wallaby comes from the Eora Aboriginal tribe, who were the original inhabitants of the Sydney, Australia area.

Wallabies can be found all across Australia, and particularly so in some of the more remote, heavily timbered, or rugged areas of the country; but less so on the great semi-arid plains that are far better suited to the larger, leaner, and more fleet-of-foot kangaroos. But it is not Australia's wallabies that we are focusing on in the pages of this book: no, it is Britain's very own wallabies – and specifically those of the county of Staffordshire – that interest us.

It is, of course, an undeniable fact that the wallaby is not indigenous to the British Isles; however, that has not stopped imported colonies from absolutely flourishing in this green and pleasant land. For example, a colony of more than 100 can be found in the Ballaugh Curraghs area of the Isle of Man - having bred originally from a single pair that escaped from the nearby Curraghs Wildlife Park some years ago. Similarly, Inchconnachan, an island that is situated in Scotland's Loch Lomond, has its own group of wallabies, which was deliberately introduced there in the 1920s by Lady Arran Colquhoun. And smaller colonies can also be found in Ashdown Forest, Sussex, and on the east-coast of England, in the county of Norfolk, and – according to a handful of stories – Suffolk, too; although the latter cannot at this stage in the game be verified.

In addition, the Centre for Fortean Zoology's Director, Jon Downes has uncovered reports of wallabies seen in the desolate wilds of Devonshire in the late 1970s, as he graphically and entertainingly relates within the pages of his warts, wild-wenches, whisky and Gonzo-driven autobiography: *Monster Hunter*. There is also a small population of such animals on Lambay Island off the east coast of Ireland, which was introduced there by Dublin Zoo after a sudden population explosion in the mid-1980s.

Then there is the story of Alice Morris, who claimed a close-encounter of the large and hopping kind in the Lake District in 1955. In this particular case, Alice says that the creature shot across the road late one night in two mighty leaps as she and her husband were returning to their then-home in the town of Oban, Scotland.

But, perhaps, the strangest of all stories surfaced in 2007, when nothing less than a fully-grown albino wallaby was seen – and photographed, no less – near Olney, Buckinghamshire, by a witness named Stacey Purdy. Interestingly, what was very probably the same animal was seen two years earlier by a man named Paul French, as he drove through woods at nearby Hanslope – only approximately five miles from where Stacey Purdy had her own close encounter of the hopping variety.

With that brief, yet tantalising, background now in hand and firmly digested, what is the true story of the wallabies of Staffordshire? Are they merely the stuff of modern-day myth, urban-legend and fantasy? The answer to that latter question is: no, they most certainly are not at all.

For many a decade, the creatures made their happy home high in the sprawling green hills of the Peak District: an upland area in central and northern England, lying mainly in northern Derbyshire, but that also covers parts of Staffordshire, Cheshire, Greater Manchester, and South and West Yorkshire, and which collectively encompasses approximately 555 square miles.

The red-necked Tasmanian wallabies in question were part of a collection of animals that had been brought to the area by a landowner named Captain Courtney Brocklehurst, of Swythamley Hall, who was killed in action in Burma during the Second World War. As far as Captain Brocklehurst's history is concerned, he was the owner of Roaches House near Buxton, where the wallabies lived – along with a variety of yaks, llamas and emus - contentedly after their transfer from Swythamley Hall, and until wartime regulations unfortunately required that his private-zoo be disposed of. No-one, of course, wanted to see the animals killed as a result of the stringent regulations; and so, as a result, they were summarily released onto the moorland – The Roaches - where they successfully survived and thrived for many a long year; as did one of the yaks, which was reportedly last seen in 1951, after which point, one presumes, it finally breathed its last on the moorlands of Staffordshire.

The Roaches (from the French Les Roches, or The Rocks) is the name given to a prominent rocky ridge situated above Leek and Tittesworth Reservoir in the Peak District, which is very popular with hikers and rock climbers. And, in clear conditions, it is possible to see much of Cheshire and views stretching as far as Snowdon in Wales and Winter Hill in Lancashire.

Certainly, sightings of the creatures on The Roaches were quite regularly reported now and again in the years and decades that followed – particularly as the colony grew in size and to the point where their numbers reached nigh-on three-figures. However, when the British Isles were hit by a particularly devastatingly-hard and notorious winter in 1962/1963, the colony's numbers were massively decimated - as a result of the animals' inability to survive the harsh English weather, of course.

Tragically, estimates suggest that the previously-thriving colony was reduced to barely six in number. However, as nature fortunately demonstrates time and time again, the colony pulled itself back from the near-catastrophic brink of overwhelming extinction, eventually increasing its figures to a healthy and respectable sixty or so by the mid-1970s.

By 2000, the situation was radically different, however, and looked very bleak, indeed. There was even widespread belief and acceptance at the time that the colony was all-but-extinct. And, in that same year, Derek Yalden, a lecturer in zoology at Manchester University, said that by the previous winter – 1999 - there were estimated to be only two females left. He added that if the creatures had not completely died out, '...they're certainly doomed. There are too many people and dogs and a lot of road casualties'.

Once again, however, nature had other, very different and grander plans in mind, and the Peak District wallabies were determined not to go down without a brave and memorable fight. As evidence of this, in July 2009, 46-year-old Andy Burton managed to capture several photographs of one of the legendary animals: blissfully hanging out on moorland between Gradbach and Lud Church.

Burton said:

> 'I used to live in Buxton years ago and heard the stories about wallabies up on the Roaches. It's fantastic to have seen one. I would like to think there are more up there,'

Bob Foster, who runs a website dedicated to the area, added:

> 'I live on the Roaches and I once saw one but that was back in 1989. People have reported wallaby sightings over the years but this is the most recent. One was spotted a couple of years ago. Everyone thought they had died out by 2000 but this shows that they haven't. Whether there is only one left we don't know.'

I am also aware of several stories and rumours suggesting that a flourishing colony of red-necked wallabies live somewhere within the deeper and denser parts of the Cannock Chase woods. To date, unfortunately, I have to admit, I have seen no evidence suggesting that such a thriving colony really does exist – at all; and all the stories relating to a large presence of wallabies on the Chase are very much apocryphal and friend-of-a-friend-like in nature.

But that doesn't make the stories any less interesting, of course. And, for the sake of completeness, I now bring them to your attention. After I had lectured one night in April 1998 at Rugeley's *Red Rose Theatre* on the subject of my book, *The FBI Files*, for the *Staffordshire UFO Group*, one audience-member came up to me and said that she was fully aware that the Cannock Chase had its very own, burgeoning population of wallabies. As I sat and listened with interest, she proceeded to tell me a wonderful tale of how the animals were said to live blissful, contented lives in one of the deeper and heavily-wooded parts of the Chase, and in a fashion that made it sound like they had just stepped (or, perhaps, bounced) out of the pages of *Watership Down*. The woman said she knew the exact location, had visited it several times, and had even gained the trust of the animals in the process!

'I know where they are, but can't tell you,' the woman maintained, in a slightly conspiratorial fashion.

Of course, this was a great story; but it was a highly unlikely one, too. I can *never* say that it contains *no* nuggets of truth at all; but, to me at least, back in 1998 it sounded very much like the work of a fantasy-prone soul – as it continues to do so today. But, who am I to loftily suggest that such a scenario is completely impossible? Perhaps, the woman's story is not the outrageous tale that it seems to be. If that *is* so, then I sincerely hope that the merry band she claimed knowledge of twelve years ago is still out there.

With all that said about literal colonies of wallabies on the Chase, however, there certainly are several apparently very credible reports from the Chase and its immediate surroundings of *solitary* wallabies having been seen bounding around the area – both during daylight hours and night-time. So, again, just maybe perhaps, there really *is* such a colony of wallabies on the Chase, after all, and one that has successfully, and astonishingly, thus far eluded all attempts to track it down. Or, maybe, these admittedly-small-in-number reports from the Cannock Chase are evidence of the occasional adventurous Peak District wallaby checking out potentially new habitats and homes.

Whatever the current status and true number of the Staffordshire wallabies, my own personal view is that – as with all of the other out-of-place creatures and critters, too that are described within the pages of this book – we should firmly embrace their presence, which surely only enhances the natural wonders of the county's wildlife and countryside.

I say: long may the wallaby call Staffordshire its home...

The Mystery Animals of Staffordshire

Glen Vaudrey

Me again, so you've read about the wallabies that Nick found and all I can add is the following reckless comment on how the wild stocks could be replenished.

It's a wonder there aren't more wallaby sighting these days considering just how cheap it is to buy these animals as pets, on occasions cheaper than the cost of a trail camera that you would need to take a shot of one in the wild. So even if this original group should disappear there is always the chance that a new wave could be just around the corner.

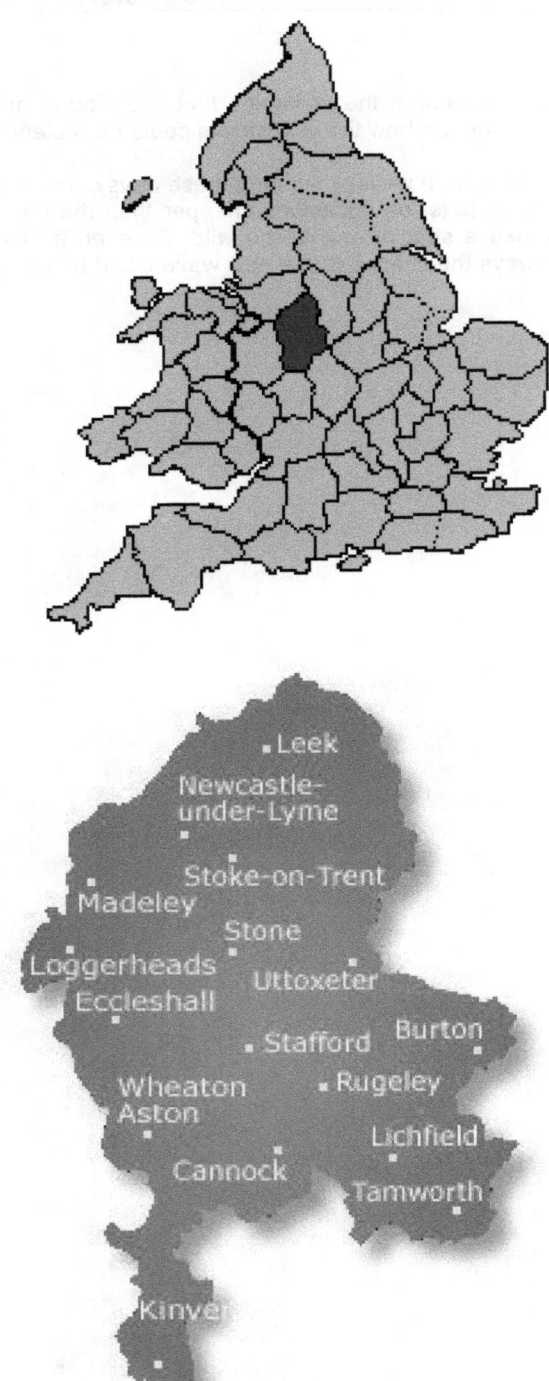

Chapter XIV
Boar to be Wild

'*...It was great to see it. It was like seeing the Loch Ness Monster. I couldn't believe it...*'

The wild boar (*Sus scrofa*) is a particular species of pig that includes no less than sixteen sub-species, and which is part of the biological family *Suidae*. It is the wild ancestor of the domestic pig, an animal with which it is able to successfully mate. These potentially-lethal and notoriously bad-tempered creatures are native across much of Northern and Central Europe, the Mediterranean, much of Asia, and as far south as Indonesia. Populations of wild-boar have also been artificially introduced in some parts of the world, most notably the Americas and Australasia; where they are hunted for both sport and food.

And as we will shortly see, populations of wild boar have also become established within the British Isles, following their escape from captivity. And, taking into consideration the mythology and folklore that surround many of the beasts referred to in this book, so too the wild boar has become the absolute stuff of legend.

In Greek mythology, two boars in particular are very well known: the Erymanthian Boar was hunted by Heracles as one of his Twelve Labours; while the Calydonian Boar was hunted in the Calydonian Hunt by dozens of other mythological heroes of times past, including some of Jason's legendary Argonauts and the huntress Atalanta. In addition, Ares, the Greek god of war, had the alleged ability to transform himself into a wild boar, and even had the gall to gore his son to death while in this particular form to prevent the young man from growing too attractive and stealing his wife.

And, from the world of Celtic mythology, we hear equally fascinating folkloric tales of the wild boar, which was sacred to the Gallic goddess, Arduinna. One such tale is that of Fionn mac Cumhaill (or Finn McCool), who lures his rival, Diarmuid Ua Duibhne, to his gored-by-boar death.

The Norse gods Freyr and Freyja both kept boars: Freyr's was named Gullinbursti – meaning Golden Mane, as a result of its unique bristles that glowed in the dark and illuminated the ancient, darkened pathways upon which Freyr roamed. Freya, meanwhile, rode the boar Hildesvini (that translates as Battle Swine), when she was not using her cat-drawn chariot.

According to the poem *Hyndluljóð*, Freyja concealed the identity of her protégé Óttar by turning him into a wild boar.

In ancient Persia (now Iran), during the Sassanid Empire, boars were respected as both fierce and brave creatures; while in Hindu mythology, the third avatar of the Lord Vishnu was Varaha; yes, you surmised correctly: a wild boar.

Swiftly moving on to ancient Rome: it is known that at least three Roman Legions had a boar as their emblems: Legio I Italica, Legio X Fretensis, and Legio XX Valeria Victrix. X Fretensis was centrally involved in the First Jewish–Roman War, which culminated with the destruction of Jerusalem and the Jewish Temple in 70 AD. In addition, it was stationed in Roman-occupied Judea for centuries and was involved in numerous other acts of oppression against the Jews. By one theory, resentment of this Legion's boar emblem, which came to be identified with extreme destruction and persecution, partly accounts for the deep-rooted traditional Jewish aversion for pork.

The wild boar is also a long-standing symbol of the city of Milan, Italy. For example, in Andrea Alciato's *Emblemata* of 1584, beneath a woodcut of the first raising of Milan's city walls, a boar is seen lifted from the excavation. Not only that: the foundation of Milan is directly credited to two Celtic peoples: the Bituriges and the Aedui, who had as their emblems a ram and a boar, respectively.

In Medieval hunting, the boar was very much a 'beast of venery', the most prestigious form of quarry, and was regularly hunted by packs of bloodhounds. And with respect to the wild-boar and hunting, a story from Nevers, which is reproduced within the *Golden Legend*, states that one night Charlemagne dreamed he was about to be killed by a wild boar during a hunt, but was rescued by the appearance of a child, who had promised to save the emperor if he would give him clothes to cover his nakedness. The Bishop of Nevers went on to interpret this dream as meaning that the child was Saint Cyricus and that he wanted the emperor to repair the roof of the Cathédrale Saint-Cyr-et-Sainte-Julitte de Nevers - which Charlemagne duly did.

Moving on to Scotland, the ancient Lowland Scottish Clan Swinton is said to have to have acquired the name Swinton for their bravery in successfully clearing their area of wild boar. Indeed, the chief's coat of arms and the clan crest allude to this legend, as does the name of the village of Swinewoo, which can be found in the county of Berwick, and which was granted to them in the Eleventh Century.

Meanwhile, in England, a Forest of Dean legend tells of a *Godzilla*-like boar, known as the Beast of Dean or the Moose-Pig, which was said to have terrorised local villagers in the early part of the 19th Century. So the story goes, in 1802, utterly tired by being harassed by the mysterious giant boar which, reputedly, was large enough to have felled trees and crushed hedges and fences, farmers from the village of Parkend undertook an expedition to capture and kill the creature, but found nothing whatsoever. The legend, however, still lives on.

So much for folklore and mythology: but what of the wild boar in Britain's real world? And: what can we learn about the presence of the beast in rural Staffordshire? Between their medieval extinction in the British Isles (or, some might very well argue, their *presumed* extinction) and the 1980s, when wild boar farming began in earnest in Britain, only a handful of captive wild boar, imported from the continent, are known to have been present in the country.

Occasional escapes of wild boar from wildlife parks *did* occur as early as the 1970s. It is since the late 1980s and early 1990s, however, that significant populations have successfully re-established themselves after escaping from farms; the number of which has greatly increased as the demand for wild boar meat has grown. Now: on to DEFRA.

The Department for Environment, Food and Rural Affairs is the arm of the British Government that is responsible for environmental protection, food production and standards, agriculture, fisheries and rural communities in the United Kingdom. Concordats set out agreed frameworks for co-operation between it and the Scottish Government, the Cabinet of the National Assembly for Wales, and with representatives from the Northern Ireland Assembly. DEFRA also leads for the UK at the EU on agricultural, fisheries and environment matters and in other international negotiations on sustainable development and climate change; although a new Department of Energy and Climate Change was created on October 3, 2008 to take over the last responsibility.

A 1998 official, governmental study of boar living wild in Britain confirmed the presence of two populations: one that roamed Kent and East Sussex and another which had made Dorset its home; both of which colonies allegedly arose as a result of damage to fences during the devastating hurricane of 1987 that allowed the animals to escape from their enclosures. Another DEFRA report, prepared in February 2008, again confirmed the existence of these two sites as 'established breeding areas' and also identified a third colony: in Gloucestershire and Herefordshire; specifically in the Forest of Dean/Ross on Wye area. A 'new breeding population' was also identified in Devon.

According to DEFRA's current estimates, the Kent-East Sussex population stands at around 200 animals; with perhaps 60 in Gloucestershire and Herefordshire, less than 50 in Dorset, and probably a similar figure in evidence throughout Devon.

Complicating the Devonshire statistics, however, is the fact that in December 2005 animal-rights activists released approximately 100 wild boars from a farm at West Anstey, Devon. Although some of the creatures *were* certainly recaptured, many continued to remain at large, with young even being sighted in the following year, some miles from the original point of release, and especially on the fringes of Exmoor. Notably, some of these beasts were filmed by local wildlife cameraman Johnny Kingdom and featured on BBC television. It has been estimated in some quarters that, today, the Devon colony may actually very well rival that of Kent and East Sussex in numbers.

As far as captive wild boars are concerned in Britain, due to the fact that the animals are included in the Dangerous Wild Animals Act of 1976, certain legal requirements have to be met prior to setting up a farm. A licence to keep boar is required from the local council, who will first appoint a specialist to inspect the premises and report back to the council. Requirements include secure accommodation and fencing, correct drainage, temperature, lighting, hygiene, ventilation and insurance.

The original British wild boar farm stock was chiefly of French origin, but from 1987 onwards, farmers have supplemented the original stock with animals of both west European and east European origin. The east European animals were imported from farm stock in Sweden because Sweden, unlike Eastern Europe, has a similar health status for pigs to that of Britain. Currently, and somewhat surprisingly, there is no central register listing all the wild boar farms in Britain; and, as a result, the total number of wild boar farms is actually quite unknown.

As for Staffordshire, although its population of wild boar is most certainly not on the scale of the well-established colonies that can be found in Kent, Devon, Dorset and East Sussex, the county is not without its notable close encounters of the boar kind.

The first Staffordshire-based case I am aware of dates from December 1987 – which, perhaps not coincidentally, is only two mere months after the well-publicised and verified escapes that followed the catastrophic hurricane of October of the same year. In this particular story, the witness, Colin Blakemore, is adamant that he saw no less than seven wild boars - three of which looked for all the world like youngsters - running at breakneck speed across low grassland near the village of Leek early one Sunday morning as he drove to a nearby town to buy a Christmas-Tree. Blakemore said that he watched the animals, fascinated and even overjoyed, by what he described as 'a smashing sight for Christmas'.

Of course, whether or not a whole group of such creatures could have successfully made the mammoth trip wholly undetected to Staffordshire from Sussex or Kent – where much of the devastation wrought by the hurricane had occurred – is a matter of keen debate. But, regardless of their actual point of origin, Colin Blakemore is in no doubt at all as to what he briefly saw twenty-three years ago. And, moving on...

In the summer of 2006, I was interviewed by Andy Richardson, a journalist with the *Walsall Express & Star* newspaper. The subject of the interview: the research I was then undertaking for this very book. During the course of the interview, I mentioned to Andy that I had received scant details of a couple of sightings on the Cannock Chase of what sounded suspiciously like wild boar.

Andy expressed amazement and proceeded to tell me how, some time previously when he had been working for the *Chase Post* newspaper, he had arrived at work one day to be told by one of the receptionists that she had seen on her way to the offices 'a big black pig run across the road in front of her'. Andy told me that, based on her description, it was quite clear that what she had actually seen was nothing less than an adult wild boar.

A fully-grown and mature wild boar, with two piglets in tow – which suggests another parent may well have been in the area, too – was seen at, of all places, the seemingly monster-friendly Castle Ring at Cannock Wood: in September 2007, by two teenage boys with an interest in archaeology, who were at the location to take photographs for a school project.

Gareth and Dean say that they saw a huge wild boar on the pathway that surrounds the Ring, just after they climbed the steps that lead from the car-park to the Ring itself. Gareth says, not without justification:

> 'It was great to see it. It was like seeing the Loch Ness Monster. I couldn't believe it.'

Dean adds:

> 'I just thought it was a big pig, at first. You know: it had escaped from a farmer or something. But when it turned around, it was definitely a wild boar. We ran right back to my dad's car and told him, because he was waiting for us. He came racing up, but we didn't see it again.'

The final case culled from my files involves not an actual sighting of a wild boar, but a distinctly boar-like grunting heard on land near the Staffordshire locale of Brockton. In this case, the date was March 2008, and the witness was walking her dog, which apparently became acutely aware of the presence of the animal, and began furiously barking at a thick area of fern in which the boar was presumably hiding. It was when the woman heard the animalistic grunting, however, that she elected to get the hell out of there as quickly as possible.

While the stories of Staffordshire's wild boar population may not be quite as hair-raising as those relative to the sightings of big cats, Bigfoot and werewolves in the county, they nevertheless serve to justify the following warning: if you ever find yourself deep within the woods of Staffordshire, tread very carefully and very quietly. Beady, vicious eyes may well be secretly sizing you up for dinner.

And, there are actually more than a few reports of other, out-of-place animals on the loose in Staffordshire that also deserve our keen attention...

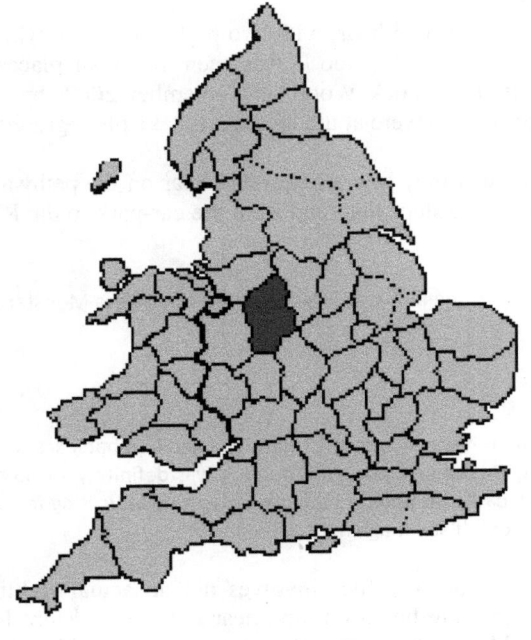

Chapter XV
Alien Animals

The coypu, or nutria (*Myocastor coypus*), is a large, herbivorous, semi-aquatic rodent, and the only member of the family *Myocastoridae*. Originally native to temperate South America, it has since been introduced to North America, Europe, Asia, and Africa, primarily by fur ranchers. Although it is still valued for its fur in some regions, its destructive feeding and burrowing behaviors make this invasive species a pest throughout most of its range.

As far as the British Isles are concerned, Coypu were introduced to East Anglia from Argentina, for their fur, in 1929. Many escaped a decade later, however - during one particularly stormy night - from the Carill-Worsley Farm in East Anglia; thus practically ensuring the inevitable establishment of a wild population of the creatures all across the region. Indeed, by 1961, none other than the prestigious *Time* magazine reported that Coypu had overrun a truly astonishing 40,000 acres in Norfolk, Suffolk and Essex and was 'munching its way inexorably northward'.

And, in 1965, even the British Government got involved in the controversy, as the following December 21 exchange of that year, extracted from *Hansard*, clearly demonstrates:

> 'Mr. Hazell asked the Minister of Agriculture, Fisheries and Food if he will make a statement about the coypu campaign which his Department has undertaken in East Anglia during the past three years.'

The reply was as follows: 'The campaign, which opened in August 1962, will end on the 31st of this month. The coypu or nutria is a South American aquatic rodent which was introduced into Great Britain about 1930 for fur farming purposes. As a result of escapes from farms coypus established themselves in Norfolk and Suffolk and spread to neighbouring counties. It became clear that this animal was capable of causing serious damage to agricultural crops, especially sugar beet and other roots, as well as undermining the banks of rivers and dykes. It was therefore decided to launch a special campaign to bring the coypu under control.

> 'Coypus have been systematically cleared by working inwards through Norfolk and North Suffolk through the heart of the infestation in the Norfolk Broads. In

the more inaccessible areas of the Broads themselves, total eradication is not possible, but even there the coypu population has been reduced to manageable limits. As a result of this campaign, farmers and other occupiers should now be able to deal with the pest as is their responsibility. My Department will advise if individual occupiers need help and rabbit clearance societies will continue to receive grant for work against 425W the coypu. The Joint Parliamentary Secretary will attend a meeting in Norwich on 6th January, 1966 to outline plans for ensuring that the pest does not reestablish itself, and he will convey my thanks to the many organisations which have contributed to the success of the campaign.'

As for the situation today, the Department for Environment, Food and Rural Affairs has stated that the coypu was successfully, and finally, eradicated in the wild in the British Isles in 1989. Or was it...?

Personally speaking, I am only aware of one, solitary sighting of a wild coypu in Staffordshire, and that was near the village of Alrewas in the summer of 1992, when a resident of Brighton was visiting friends in the area and saw what she thought were 'two fat guinea-pigs running up the road'. They were not overweight guinea-pigs, however. Attempts by the witness to identify the creatures in question eventually satisfied her that what she had seen was most definitely a pair of mature, fully-grown coypu. Perhaps the animal, despite the very best efforts and the assurances of DEFRA, continues to maintain a foothold in Britain and right in the heart of Staffordshire - albeit, I have to strongly suspect a highly precarious foothold at that.

Porcupines are rodents with a coat of sharp spines, or quills that help defend them from predators. They are endemic in both the Old World and the New World and are the third largest of the rodents, behind the capybara and the beaver. Porcupines come in various shades of brown, grey, and the unusual white. The name itself is derived itself from the Middle-French porc d'épine which can be translated as 'thorny porc', 'spined porc' or 'quilled porc', hence the nickname 'quill pig' for the animal. Very appropriately, a group of porcupines is called a 'prickle'.

Tales of porcupines on the loose in the British Isles are not at all widespread; but they do exist. For example, one was at large in the Forest of Dean in 2005, after escaping from a farm at Elwood, near Coleford, where it was temporarily being housed, and while awaiting transfer to a zoo in Yorkshire.

In Staffordshire, the situation is equally as scant; however, we do have one case on record: namely, of two porcupines escaping from a county-zoo at some point during the 1970s. Whether the animals survived and bred, however, we do not really know. It would, however, be very satisfying to think that somewhere, deep within the Staffordshire greenery, lives a happy, contented and frolicking little colony of the creatures; blissfully unaware of what their discovery might mean at a regional, zoological and even cryptozoological level.

The Mystery Animals of Staffordshire

Armadillos are small placental mammals, known for having a leathery armor shell, and whose name is Spanish for 'little armored one'. There are approximately 10 extant genera and around 20 extant species of armadillo, some of which are distinguished by the number of bands on their armor. Their average length is about 75 centimeters (30 in), including tail; however, the Giant Armadillo can grows up to five-feet in length, and may weigh-in at 130 lbs.

In the United States, the sole resident armadillo is the nine-banded armadillo (*Dasypus novemcinctus*), which is most common in the central southernmost states, particularly in Texas, where I have made my home for the last decade. Their range is as far east as South Carolina and Florida and as far north as Nebraska; and have been consistently expanding their range over the last century due to a lack of natural predators and have been found as far north as Illinois, Indiana and southern Ontario.

But, how do we even begin to explain a solitary sighting of an armadillo seen on a main-road near the city of Lichfield in the mid-1960s? Well, the answer, of course, is that we *can't* explain it – at all. The case is long-gone and very much friend-of-a-friend-like in nature. In addition, the creature itself would, without doubt, be long-dead by now, and even the single witness has now shuffled off of this mortal-coil, too. However, Muriel Atkins says that she heard the bare bones of the story around 1970 or 1971 from a friend and colleague while working at a furniture shop in Lichfield. Muriel says today that:

> 'I forgot about it for years; but then [I] started telling people a few years ago when all those black-panther stories started to pop up. I suppose it was similar, really.'

On January 13, 2010, the *Stafford Post* newspaper ran a story titled *Snow May Solve the Pine Marten Puzzle* that went as follows:

> 'Wildlife experts hope the bitter weather will at last solve the mystery over whether one of Britain's rarest and most elusive mammals has a paw-hold in our area. The Vincent Wildlife Trust is calling on Chase residents to watch out for pine martens - large, stoat-like creatures - which may be lured into the open in the search for food. For many years, it was believed the cat-sized animal existed only in the wilds of Scotland and a pocket of mid-Wales. Then reports of sightings began to come-in from all over the country - including 17 in our area.'

The newspaper continued:

> 'Our own Staffordshire Wildlife Trust sat up and took notice three years ago when a pine marten was seen in Hednesford. Head of the county trust, Derek Crawley, believes that animal may have escaped. However, the county trust has now placed 17 pine marten boxes in secret locations - and Derek says he wouldn't be surprised if the creatures were on Cannock Chase.'

The 'Post' quoted Crawley's words on the matter:

'I believe that population would be very low and spread out, but I wouldn't be surprised if they're here. The truth is we don't know how many are out there and where they are in the country. Pine martens are extremely secretive - you've got about the same chance of seeing one as an otter. They are predated on by foxes, so they seek refuge in rocky outcrops, such as quarries, and large trees. We want members of the public to look out for distinctive five-toed prints in the snow. Details of actual sightings or pictorial evidence would be even better.'

The *Stafford Post* expanded further:

'In total, there have been 70 reports of pine martens from the West Midlands over the last 12 years. They are from Herefordshire 25, Shropshire 22, Staffordshire 17, Worcestershire six, Warwickshire one. Neil Jordan, pine marten project officer at the Vincent Wildlife Trust, said: "There are probably fewer than 10 individuals in the West Midlands now, although they are likely to avoid contact with humans, so it is difficult to be sure. We are very anxious for local help in building up an accurate picture. We are hoping that in the cold weather one or two might venture out." Pine martens are about the size of a domestic cat, have a bushy tail and distinctive cream bib. They live in ancient woodland and nest high in tree cavities.'

Only forty-eight hours later, the *Chase Post* newspaper stated the following under the title of *Cannock Chase is hit by Muntjac Mayhem*:

'A new and unwelcome visitor is causing a headache for those who tend Cannock Chase woodland. The number of tiny Muntjac deer - only fractionally taller than a dog - has risen sharply in recent years. And that's a concern for the Forestry Commission, which fears the grazing animals may play havoc by destroying saplings and woodland flowers. They've already decimated the carpets of bluebells at Bagot's Wood, Abbots Bromley - believed to be the source of the Chase population explosion.

'Commission operations manager Gordon Wyatt admitted: "They are an unwelcome visitor. They've only been on the Chase for about a decade, but we are seeing a lot more of them now. They're an introduced species and breed all year round, so populations can rise rapidly."

'Muntjac - no taller than 52cm - have come a long way since being introduced from their native China to Woburn Park, Bedfordshire, in the early 20th century. Deliberate releases plus escapes quickly led to a feral population, with the local hotspot being Bagot's Wood. Their rise in numbers adds to the success of the Chase's overall deer population. There are now over 1,000 on the Chase - the vast majority being fallow, though red deer live on the fringes. To keep numbers down, a cull of around 200 takes place every year.

'And Gordon stressed the animals are coping very well indeed with the freezing conditions. "We don't put food out for them. The forest floor is free of snow and there is still plenty of food available to them."'

And, in view of all the above, we surely have to wonder what else of a wild and animalistic nature has decided to call Staffordshire its home…

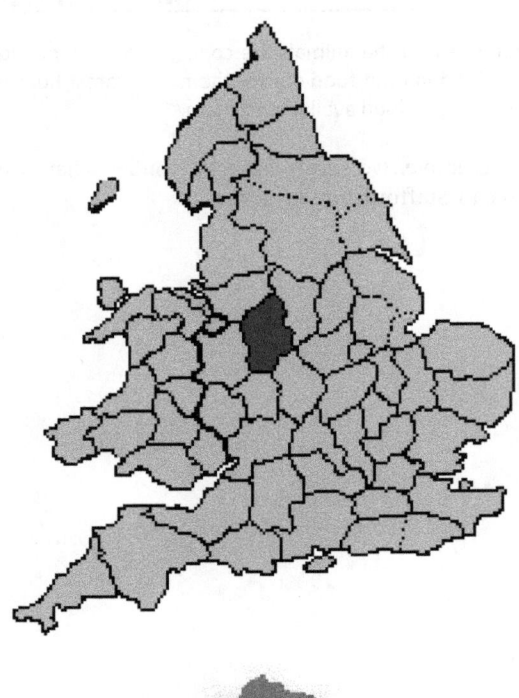

Chapter XVI
Flying Monsters
'...It reminded me of a devil...'

Within the realm of monster-hunting, sightings of large, winged beasts of unknown origin absolutely abound. Throughout the course of my time as an investigator of the unexplained, I have encountered numerous such cases, which collectively suggest that the skies of our world are indeed populated by monstrous entities of unknown origins. And, several of them can be found in Staffordshire, too – such as the devilish entity seen by Pauline Charlesworth in 1986, as described in Chapter 8 of this book, *Strange Creatures of the Castle Ring*.

But, before we dig much further into such matters within Staffordshire, it's worth noting the rich and varied history of this particular aspect of cryptozoology – as well as a few, classic cases, too.

In the early part of 1946, at an aged and sinister-looking house that at the time was situated upon the fringes of a small and overwhelming isolated Texas town, specifically just outside of the city of Lubbock, deep and dark events were brewing. According to a local legend, on one occasion in the dead of night, a group of kids playing in the area had seen two eight-foot-tall humanoid creatures climb stealthily out of the building's cellar.

The creatures were not only eight-feet-tall: they were also gray of skin, had large, leathery wings, and glowing red eyes. The monstrous pair apparently turned sharply as they surfaced from their underground lair and stared intently at the kids, then broke into a hopping-style run, opened their immense wings and soared majestically into the starlit sky. One interesting observation was that the limbs of the creatures looked almost hollow against the background of the full moon that loomed overhead. Supposedly, one of the original two creatures was seen several months later, standing in the middle of the local highway, by a terrified motorist in the early hours of the morning and while issuing a woeful moan.

Well-known Kent-based cryptozoologist, researcher and author of several acclaimed books on mystery animals Neil Arnold has dug into a now-renowned event of a distinctly similar nature from Britain, and that took place at Sandling Park, Hythe in 1963. Neil recorded that:

> 'Four teenagers, one being seventeen-year-old John Flaxton, saw a strange light over the park late one night as they returned home from a party, and then ran in horror from a creature that emerged from nearby woodlands where they swore the light had landed. For several days after the event the area was bathed in an eerie glow. No-one ever mentioned the creature flying, but what it was remains a mystery.'

Neil added that:

> 'Local UFO experts believed that the case was nothing more than a misinterpretation of natural phenomena, but Flaxton recalled: "I felt cold all over," while another witness, eighteen-year-old Mervyn Hutchinson, told police: "It didn't seem to have any head. There were huge wings on its back, like bat wings."'

And there was even more to follow, as Neil graphically detailed:

> 'On 21 November, that same year, seventeen-year-old Keith Croucher claimed also to have seen a weird craft in the area, this time floating over a football pitch near the park. Two days later John McGoldrick went to the area with a friend to look into the weird reports and claimed to have discovered an area of bracken as if something disc-shaped had landed there. Three giant footprints were also found in the vicinity which were said to have measured two-feet long and nine inches across. On 11 December, various newspaper reporters accompanied McGoldrick to the area and found that the woods were illuminated by an eerie, glowing light. No-one investigated any further and the case faded as mysteriously as it had emerged.'

A vile, winged nightmare equipped with a pair of glowing, red eyes, the Mothman of Point Pleasant, West Virginia is undoubtedly the most infamous of all the many and varied sky-beasts that have made a name for themselves within the annals of cryptozoology. Its mid-1960s manifestations ominously coincided with a massive wave of UFO incidents, encounters with the dreaded Men in Black, and a whole range of Fortean high-strangeness of a mind-boggling nature – all of which chose to descend upon the unfortunate city, and the people too, of Point Pleasant.

The bizarre series of events came to an absolute climax on December 15, 1967, when the city's Silver Bridge that crossed the Ohio River and connected Point Pleasant to Gallipolis, Ohio, broke away from its moorings and plunged into the river, tragically taking with it nearly fifty lives. And although a down-to-earth explanation was most definitely in evidence – that a problem with a single eye-bar in a suspension chain was to blame – many took the view, and still to this day continue to take the view, that the Mothman was behind it all.

And there we have the Owlman. In the hot summer of 1976 the trees surrounding Mawnan Old Church, Cornwall, England became a complete hotbed of monstrous goings-on, when what is surely the closest thing that Britain has to the aforementioned Mothman first surfaced. Most of those that encountered the beast reported that it was human-like in both size and shape, and possessed a pair of large wings, bright-red eyes, distinct claws, and exuded an atmosphere that reeked of dread and terror.

One of the first to see the Owlman, a young girl named Jane Greenwood, wrote a letter to the local newspaper, the *Falmouth Packet* during that same summer that detailed her terrifying experience with the creature:

> 'I am on holiday in Cornwall with my sister and our mother. I, too, have seen a big bird-thing. It was Sunday morning, and the place was in the trees near Mawnan Church, above the rocky beach. It was in the trees standing like a full-grown man, but the legs bent backwards like a bird's. It saw us, and quickly jumped up and rose straight up through the trees. How could it rise up like that?'

Notably, when Jon Downes and I journeyed to Puerto Rico in 2004 in search of the blood-sucking Chupacabras, we spent several hours of one particular day interviewing a woman named Norka who lived high in the hills of the El Yunque rain-forest, and who had also an encounter with an Owlman-style entity in the mid-1970s.

Norka explained that she was driving home one night when she was both startled and horrified by the shocking sight of a bizarre creature shambling across the road. She described the animal as being approximately four feet in height, and having a monkey-like body that was covered in dark brown hair or fur, wings that were a cross between those of a bat and a bird, and glowing eyes that bulged alarmingly from a bat-style visage. Elongated fingers, with sharp claws that looked like they could inflict serious damage, flicked ominously in Norka's direction. She could only sit and stare as the beast then turned its back on her and rose slowly into the sky.

We also have the Pterodactobats of the United States – a very apt term for such entities that was created by Neil Arnold, and who says:

> 'Tribes often describe creatures that died out millions of years ago; huge, beaked, Pterodactyl-like flyers, but surely they are mistaken. If they are wrong, then what they are seeing are either huge, undiscovered species of bat, or extremely weird manifestations that make no sense at all. During the 1970s several reports came from the Rio Grande area, where rancher Joe Suarez claimed that something, which left no tracks, had ripped his goats to shreds. However, on the night of the attack, Mr. Suarez claimed to have heard the eerie sound of large wings flapping.'

Neil says further:

> 'Such strange, flying creatures seem to flit between folklore and reality of natives and are rarely followed up by zoologists and researchers, and so often doubted until discovered by accident. Pterodactobats remain as that confused, unidentified species of membranous-winged, beaked, screeching zenith invader that should have died out over one hundred million years ago, but which still linger in the remote swamps of the world as ghosts embedded in the minds of the locals, to the extent that they are a much feared spiritual beast.'

On March 15, 1985, the *Ealing Gazette* newspaper printed a feature with the title of *Flight of the Phoenix*, and that detailed the then-recent reports of a large winged creature seen soaring over Brentford – an area of west London. The newspaper article stated:

> 'Is it a bird...? Or is it a plane...? No, but it could be a Griffin! The bizarre giant bird has been seen several times – and some people believe it could be the legendary half-eagle, half-lion returning to the area where it has forged many historic links.'

The *Gazette* focused its story upon the account of a man named Kevin Chippendale, who stated that:

> 'I saw it twice in the same place about 30 feet up by the Green Dragon flats. The shape of the bird is what attracted me in the first place. When I first saw it I thought it was an airplane. It had a very definite shape.'

In the immediate wake of the publicity afforded the sighting of Kevin Chippendale, additional tales of huge, flying animals seen in the area began to surface. In addition, the well-known author Andrew Collins wrote a small, self-published report on the matter that was as unbiased as it was informative.

A large percentage of Collins' booklet centered upon Kevin Chippendale's story, and who advised Collins that the creature he viewed was around 'the size of a large dog'.

He additionally reported:

> 'It had four legs and what appeared to be paws. There was no beak...its skin surface seemed smooth, not feathery, and it did look aerodynamic. The wings did not seem to be flapping like those of a normal bird in flight. They were much slower, almost as if they were moving in slow motion.'

A quarter of a century later, the Brentford Griffin is still discussed in hushed tones by those who still recall that strange and unnerving period of high-strangeness.

And with that background on the winged wonders of the world firmly fixed in our minds, let us move a little closer to home: to central England, of course.

The Mystery Animals of Staffordshire

The following case is not a new one. But, it is, perhaps, the most fascinating, and creepiest, of all to burst forth out of the darkened, shadowy woods of Staffordshire. The source of the story is one 82-year-old Alfred Tipton, and the location: Needwood Forest, a once-sprawling area of ancient woodland in Staffordshire which is now mostly, and tragically, destroyed.

Needwood Forest was a chase, or a royal forest, that was given to Henry III's son, Edmund Crouchback, the 1st Earl of Lancaster, in 1266, and was owned by the Duchy of Lancaster until it passed into the possession of Henry IV. In 1776, Francis Noel Clarke Mundy privately published a book of poetry called *Needwood Forest* which contained his own poem of the same name and supportive contributions from Sir Brooke Boothby Bt., Erasmus Darwin and Anna Seward. Anna Seward regarded this poem as 'one of the most beautiful local poems'. The purpose of Mundy's poems was to resist calls for the enclosure of the forest. Anna, Seward herself wrote a poem called *The fall of Needwood Forest*. Under an enclosure act of 1803, however, commissioners were allowed to deforest it, and by 1811 the land had been divided amongst a number of claimants.

In 1851, Needwood Forest was described as forming 'one of the most beautiful and highly cultivated territories in the honour of Tutbury, which contains 9,437 acres of land, in the five parishes of Hanbury, Tutbury, Tatenhill, Yoxall, and Rolleston, and subdivided into the four wards of Tutbury, Barton, Marchington, and Yoxall, which together form a district of over seven miles in length and three in breadth, extending northwards from Wichnor to Marchington Woodlands'.

Today, however, things are sadly *very* different, and most of the ancient woodland is now tragically gone: presently, the area is comprised of twenty farms, on which dairy farming is the principal enterprise; and less than 500 acres of woodland now remain. Some parts of the forest are still open to the public, including Jackson Bank: a mature, mixed 80-acre area of woodland which can be found at Hoar Cross near Burton upon Trent and which is owned by the Duchy of Lancaster.

And then there is Bagot's Wood near Abbots Bromley, which claims to be the largest remaining part of Needwood Forest, and which takes its name from the Bagot family, seated for centuries at Staffordshire's Blithfield Hall.

Situated some 9 miles east of Stafford and 5 miles north of Rugeley, the Hall, with its embattled towers and walls, has been the home of the Bagot family since the late 14th Century; while the present house is mainly Elizabethan, with a Gothic façade added in the 1820s to a design probably by John Buckler.

In 1945 the Hall, then in a neglected and dilapidated state, was sold by Gerald Bagot, (the 5th Baron Bagot) together with its 650-acre estate to the South Staffordshire Waterworks Company, whose intention was to build a reservoir, and which was completed in 1953. The 5th Baron died in 1946, having sold many of the contents of the house. His successor and cousin, Caryl Bagot, repurchased the property and 30 acres of land from the water company and began an extensive programme of both renovation and restoration.

The 6th Baron died in 1961 and bequeathed the property to his widow: Nancy, Lady Bagot. In 1986, the Hall was divided into four separate houses, the main part of which incorporates the Great Hall and is owned by the Bagot Jewitt Trust. Lady Bagot and the Bagot Jewitt family remain in residence.

And, it is against this backdrop of ancient woodland and historic and huge old halls that something decidedly strange occurred back in the summer of 1937, when Alfred Tipton was but a ten-year-old boy. And like most adventurous kids, young Alfred enjoyed playing near Blithfield Hall, and in the Bagot's Wood, with his friends: on weekends and during the seemingly-never-ending school-holidays. And, it was during the summer holidays of 1937 that something strange and monstrous was seen in that small, yet eerie, area of old woodland.

According to Tipton, on one particular morning he and four of his friends had been playing in the woods for several hours and were taking a break, sitting on the warm, dry grass, and soaking in the sun. Suddenly, says Tipton, they heard a shrill screeching sound that was coming from the trees directly above them. As they craned their necks to look directly upwards, the five pals were horrified by the sight of a large, black beast sitting on its haunches in one particularly tall and very old tree, and 'shaking the branch up and down with its claws tightened around it'. But this was no mere large bird, however.

Tipton says that 'it reminded me of a devil: I still don't forget things and that is what I say it looked like'. He adds that the creature peered down at the five of them for a few moments and then suddenly opened up its large and shiny wings, which were easily a combined twelve-feet across, and took to the skies in a fashion that could be accurately described as part-flying and part-gliding, before being forever lost to sight after perhaps 15 or 20 seconds or so.

Significantly, when shown various pictures, photographs and drawings of a wide variety of large-winged creatures that either still roam our skies or did so in the past, the one that Tipton said most resembled the creature he and his mates saw was a pterodactyl. Of course, the pterodactyl is *long* extinct; however, Tipton is adamant that the beast the boys encountered was extremely similar to the legendary winged monster of the distant past.

Were the boys merely spooked and confused by their sighting of a large, exotic bird – albeit one of a conventional nature and origin, and perhaps even a circus- or zoo-escapee? Or, was some hideous winged-thing really haunting Bagot's Wood on that fateful, long-gone morning back in 1937? Sadly, probably neither we nor Alfred Tipton will ever know the answers to those thought-provoking and controversial questions.

At around 9.30 p.m. on the night of Monday, March 8, 2004, Steve Nicklin and a friend were 'walking across Northicote Farm' near the town of Wolverhampton – which is only a very short drive from Staffordshire's Cannock Chase - when, as they neared an old

Tudor-era farmhouse they were shocked rigid by the sight of a horrifying creature that was perched within the branches of a nearby, large pine tree. Nicklin later stated that it was human in shape, 'grayish' in colour, and in excess of seven feet in height. Of relevance to the data contained in this chapter, he added:

> 'It had two legs and two arms connected to membrane-type wings. Its clawed arms seemed connected to these wings; a bit like a pterosaur.'

Nicklin expanded in astonishing fashion:

> 'It moved its head and looked directly at us, since the moon was full and there were street lights not far away. It turned its head from us, took one giant leap and glided to the next tree. The tree bent under this creature's weight as it took the impact. It then turned its head to look at us once more. Its gaze felt that it could look into our very being. We both felt quite scared. We both fled quickly. We sat down and discussed what we had seen. It was a creature the like of which we had never seen before, nor want to ever see again.'

Interestingly, its extreme height aside, the beast seen by Nicklin and his friend sounds not at all unlike that encountered by Pauline Charlesworth at the Cannock Chase's Castle Ring in 1986. And, in addition to that intriguing fact, we have the following from the *Paranormal Awakening* group – whose activities are cited in Chapter 8 – that was reported by the *Chase Post* newspaper to have uncovered details of 'an eerie, large floating/flying shape with red eyes' seen near the towns of Brownhills and Great Wyrley; both of which are situated just a few miles from the Cannock Chase.

A flying shape with red eyes: could this, possibly, have been the same creature – or one of the same species, at least - seen by Pauline Charlesworth back in 1986 and by Steve Nicklin in 2004? Perhaps we should not rashly dismiss such a possibility out-of-hand.

What was without any shadow of a doubt the strangest of all stories of unidentified flying entities seen in the skies of Staffordshire surfaced on February 19, 2009, when *Chase Post* editor Mike Lockley stated that nothing less than a flying man had been seen soaring over and around the Cannock Chase!

'Five locals have contacted the 'Post' after witnessing the figure travelling, seemingly unaided, over houses at around 11am on Sunday, February 8. One described it as a "Superman" moment - a clear case of "to Chadsmoor and beyond,"' said the newspaper.

Lockley added:

> 'But eagle-eyed Boney Hay villager Clive Wright believes those who reckon they witnessed something supernatural are talking a load of kryptonite. The 68-year-old, who spotted the flying man from the living room window of his Sunnymead Road home, believes the pilot was travelling with the aid of a jet pack - a strap-on engine made famous in the 1965 James Bond movie,

Thunderball.

'Clive's wife, Janet, 68, and 14-year-old grandson Nicholas also witnessed the Chase's own rocket man. Clive said: "To say it was strange would be an understatement. And the 'bottle' didn't come into it because none of us drink. At first I was watching quite a number of seagulls and noticed what I thought was one in the middle moving quite slowly. I got up to take a closer look and realised it was a flying man. I searched the sky for the plane he had baled out of, but could see nothing."'

The 'Post' additionally quoted Clive Wright as saying:

'All I could see was this man travelling in a controlled, straight line, travelling from Ryecroft shops across to Gentleshaw Common. I immediately went upstairs to get my binoculars and went out on the backyard, but he was gone. Some kind of Dan Dare spaceman - that's what it looked like. The only explanation is that he was wearing a jet pack, although I was surprised he was travelling over a densely populated area.'

Whether an intrepid flyer equipped with a startlingly high-tech piece of aerial gadgetry or not, the mystery of Staffordshire's flying man remains.

On January 14, 2010, the *Tamworth Herald* newspaper published a fascinating story that concerned the appearance in Staffordshire of a winged entity that had a truly disturbing history attached to it. The newspaper recorded that:

'A bat-like moth – said to be the harbinger of death – found its way from South America to a Tamworth home after being discovered in a crate of beer. Lyndsey Walker, of Wordsworth Ave, Leyfields, said her father discovered the moth at a distribution depot in North Warwickshire. And the creature has been identified as the Black Witch moth, the largest moth north of Mexico, which can grow to have a wingspan of six inches.'

The 'Herald' added the following:

'Mythology says that of one of these moths flying into the house is bad luck, often being associated with the death of a sick person. And in some parts of Mexico people joke that if one flies over someone's head, then the person will lose their hair. "I'm glad this one was dead," Lyndsey told the *Herald* this week. "The moth had been brought over in a shipment of beer. It had travelled for six weeks." She added: "My dad did not know what it was. He brought it home in his lunch box. I could not believe it when I saw it."'

And the 'Herald' was still not quite done with the story:

'Lyndsey said she had shown the moth off to children at Lark Hall Infants and that her two children, Tia (11) and Tamie (9) were planning to do a talk about the insect at their school, Flax Hill Juniors.

'In Jamaica, under the name "duppy bat", the moth is seen as the embodiment of a lost soul or a soul not at rest – but some say the moth is the embodiment of the soul of a loved one returning to say goodbye. "It is huge – I have never even seen a bat that size. It is like a bird," Lyndsey continued. "The children have been fascinated by it. It's a good specimen but is quite delicate."'

As of all of the above cases serve to demonstrate, within the skies of Staffordshire, truly unusual, unsettling and sometimes downright uncanny flying beasts lurk...

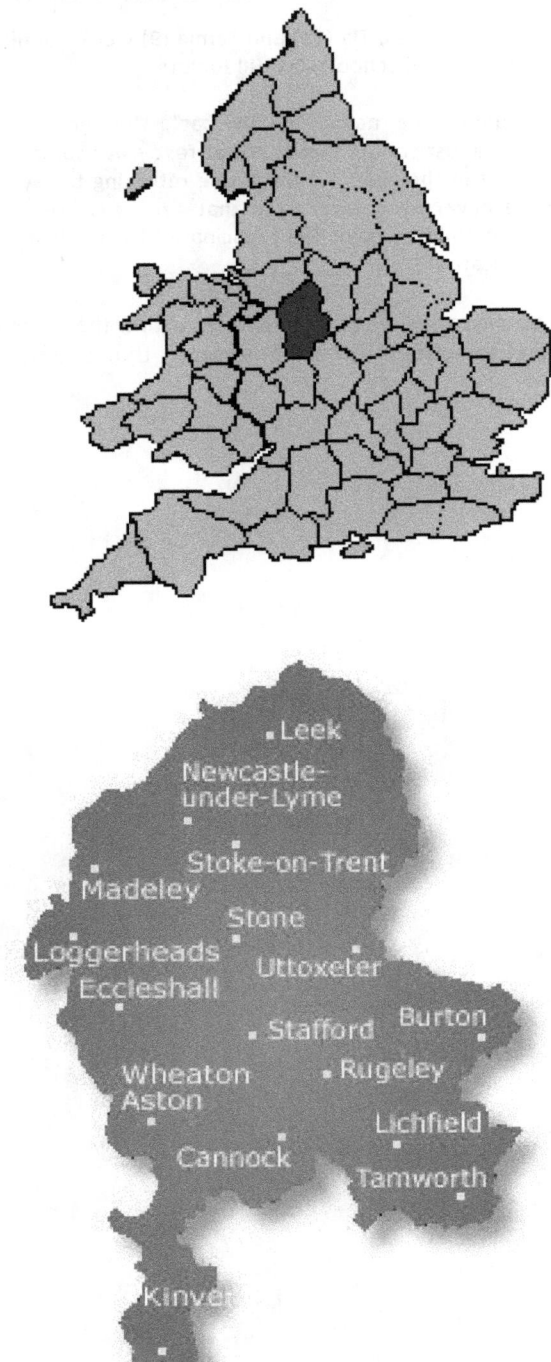

Chapter XVII
Bad Bunnies
'...it was only three days later that his father died...'

At long last Glen's very own chapter!

Having read about phantom black dogs, Jenny Greenteeth and werewolves you might be surprised to find that there is yet another big nasty creature that lurks in the folklore of Staffordshire. This beast is truly terrifying for it is none other than the ghostly rabbit of doom. Actually it's not just one rabbit of doom, it appears that there are a couple of them on the loose.

The first rabbit is one that follows in the tradition of many ghostly creatures fated to mark the death of those unfortunate enough to see them. Depending whereabouts you are in the country the doom-hinting visitation could be anything from seeing a ghostly funeral procession, a phantom lady or an animal such as a phantom dog. However in Staffordshire there was one other ghostly animal that foretold of tragic events and that was the humble bunny.

This unlikely harbinger of despair is reported as being a white rabbit that is said to haunt Kidsgrove. Yes, that same Kidsgrove that has the boggart, the dog and the headless woman.

This creature is said to have been sighted in the avenue leading to Clough Hall and in true animal of doom tradition its sighting is said to predict a death. Those who do spot the bunny may be glad to know that those who appear to die from the sightings are not the witnesses themselves but rather someone they know, okay maybe that information might not be a great comfort, I suppose it depends how you get on with the in-laws. According to one witness, whose identity is long lost in the depths of time, after he had seen the bunny of death it was only three days later that his father died. While other equally anonymous sources have been recorded that within days or weeks (and I bet years) of a sighting they have known of relatives, friends and come to think of complete strangers die. What more proof do you need of the power of the dread stare of the bunny of death? As you may have guessed I am not a great believer in the idea that a rabbit could really be the harbinger of doom, let's face it any creature that looks so cute and cuddly is not one to easily instil fear.

Our second ghostly rabbit does not seem to be foreteller of a death but rather marks the events of a fateful day long ago, this rabbit appears at the scene of a shocking murder. This is the white rabbit of Etruria.

While Etruria might sound like a rather exotic place it is actually a suburb of Stoke-on-Trent, it was the penultimate site of the Wedgewood pottery business with the name coming from the Italian district Etruria, which is good distance from Staffordshire. The original Etruria was the home to the Etruscans, a pre-Roman people who were known for both their metalwork and their terracotta sculpture. They were not, it seems however, known for their ghostly rabbits.

Fortunately there are other places closer to home known for their ghostly murder-produced rabbits. The nearest place to Etruria is in Lancashire in the delightfully named village of Crank. The Crank rabbit's first appearance dates back to the seventeenth century and involves a nameless old woman who occupied a hovel within the hamlet which she shared with her daughter Jenny and Jenny's pet rabbit. Fitting for the time the old woman was soon branded a witch by a neighbouring farmer by the name of Pullen who was convinced that he had been bewitched by her. Local belief had it that by drawing blood from the witch the spell would be broken and it was to this end that Pullen and his accomplice Dick Piers broke into the old woman's cottage. Rather clumsily they awoke Jenny with the racket as they attempted to cut the old woman's arm, not wanting to hang around Jenny ran off into the night taking her rabbit with her. She was pursued by Piers and while she managed to evade capture she succumbed to exhaustion and was found dead in the chapel. Her white rabbit didn't fare much better as it was beaten to death. Of course just because you've beaten the bunny to pulp doesn't mean that it won't come after you.

It was only a month later when the white rabbit appeared and started to follow Piers around the hamlet eventually being held responsible for driving him to commit suicide in Billinge Hill quarry. But the vengeful bunny wasn't finished yet as it now set about hounding Pullen who was to be seen being pursued across the fields until he died frightened and exhausted by the chase.

So there you have a brief recap of the powers of a vengeful bunny, there are many elements of the story that could suggest that it is nothing more than a story designed to entertain. The tale from Etruria however certainly has its feet set in historical fact; we need to travel back to a Saturday afternoon in August 1833 when John Oldcroft was murdered.

The following details are from an engraving from a nineteenth century woodcut published by G. Smeeton, Printer, 74, Tooley Street. 1834

'The trials of Charles Shaw, aged 16 for murdering John Oldcroft, aged 9.

Charles Shaw
A well-looking lad of 16 was charged with the wilful murder of Charles[sic] Oldcroft, aged 9, by the fixing of a cord round his neck. The only assignable motive of the prisoner was the commission of the offence was the acquisition of 1s. 6d. which the deceased possessed. The prisoner and the deceased were both in the service of Mr. Hawley, a potter. Both had been paid their wages at about

six o'clock, on 3d of August last. The deceased had due to him 1s. 4½d; but, as he was a good lad his master made it up to 1s. 6d. in a few minutes after they were paid, the deceased and the prisoner went away together, they were seen by various persons going towards the Etruria race-course. On their way they passed a boy named Robinson, who was bathing in the canal. He saw the prisoner dangling a piece of cord, and when afterwards the prisoner was reminded by Robinson that such was the case, he denied that he had any cord at all. The prisoner and the deceased, after some little rough play with Robinson, then went on in the direction of Maccaroni-bridge, and they were seen playing with copper money on the bank. The deceased not returning, his parents went in search of him, and proceeded to the house at which the prisoner lived, and the account he gave was, that he had left the deceased with gamblers, at the race-course. The parents continued their search without success; but one of the peasantry having a wood-lark, which is a bird fond of some of the contents of a wasps' nest, he went in search of a wasps' nest, and found concealed under some willows the body of the deceased, with a cord round the neck, corresponding in appearance with that that the prisoner carried. In consequence of suspicion the prisoner was taken up. There were several marks of blood on the prisoner's shirt, which had been covered with potter's clay, but it was found to be blood, he said his nose had bled, and he began rubbing his nose as if to make it bleed. Evidence was also given that the prisoner had told another lad in prison that he killed the deceased. The jury returned a Verdict of Guilty, and Mr. Justice Patterson ordered him for execution on the Wednesday following March 19th.'

A shocking crime you will agree, no wonder it is held to be the reason behind the appearance of the white rabbit of Etruria. In the years following the crime there started to appear stories of something very odd happening. Back in the early nineteenth century for folk wanting to get between Etruria to Hanley there was no road only a path that went across the fields. The stretch between Etruria to Cobridge was between closely placed trees that overhung the path forming a canopy over the top of it, it was hardly surprising then that come night time this route took on a slightly eerie feel to it. You might think that it would be quiet at night but that isn't always the case, there are always strange noises to be heard, hedgehogs sneezing, cats wailing, owls hooting, these could be the usual suspects, but travellers along that path got something completely different, the piercing screams of a boy in great distress screaming out for help (I bet that resulted in a few changes of trousers), just as suddenly as it started the noise would stop and a deathly still quiet would descend on the scene, at that moment a milk white rabbit would appear. It would jump out from one side of the pathway and run along the track for a short while before it would just disappear. It was said that this ghostly white bunny would always follow the same route.

These unearthly goings on seemed to occur always in the same place known as The Grove, a place that steadily became best avoided as fear of this ghostly bunny grew. Of course there are always some who think they know better and it was such in this case. The gentleman in question whose name appears lost in the depths of time decided that he was going to catch the rabbit. He went to The Grove and awaited the hour of the rabbit to arrive. As if by clockwork come the hour come the screams for mercy followed by a deadly unnerving silence, then suddenly the white rabbit appears and starts off down the path, as it turns out to where the man was waiting for it, with a leap the man goes to grab the beast, he is certain that he managed to grab the bunny, however all is not as it seems as the rabbit had vanished. All he seemed to get for his trouble was a dislocated shoulder, tough luck but at least his run in with the white rabbit didn't end in members of his family tripping of this mortal coil or even random folk he knew going off to meet their maker.

It seems reports of the ghostly white rabbit have faded over time and certainly there does not appear to be a record of more recent sightings. Perhaps the shocking events of that time have now faded so much from thought that a strange noise in the area will no longer be assumed to be that of person screaming for mercy in the night. Or perhaps the bunny found somewhere else to haunt. Or did it go tracking down the killer? You see while Charles Shaw was sentenced to death his fate was not to end his days dangling from a rope, rather his sentence was commuted due to his age and he was transported to Van Diemen's Land (today known as Tasmania), leaving England on the *Norfolk* in May 1835 and arriving in August 1835. It seems he got on with his life, deciding to stay there and in 1850 marrying an Amelia McCabe, after that who knows what happened to him, I am sure someone out there does.

Before I close the file on the bunnies of doom let's speculate on the reason the white rabbit of Etruria vanished. Could it simply be the case that belief in such a creature disappeared as the area became more built up, or could it be that the rabbit was really a vengeful creature that set off following the guilty man to see that justice was done. Perhaps it just simply set off after him not knowing how long it would take a rabbit to travel hopping along all the way to Van Diemen's Land.

Perhaps the final earthly sound that Charles Shaw heard was the munching of a carrot and the immortal line

'What's up doc?'

Conclusions

And, now, our long and winding journey around the wilder parts – and, sometimes, around even the not so wild parts - of Staffordshire, in search of unknown animals and out-of-place beasts, is finally at its very end. Having carefully discussed, dissected and digested an absolute plethora of strange tales on what can only be described as a true menagerie of strange creatures indeed, what, if anything at all, can be said with any degree of certainty about the mystery animals of the fair and green county of Staffordshire, England?

Well, first of all, it's very clear that in many cases we are simply dealing with exotic animals that – whether by accident or design, or maybe, perhaps, even as a result of a bit of both – have no real place in Staffordshire, yet that are seemingly content to make the county their long-term home. I'm talking here, of course, about the county's resident populations of wallabies, wild boar, and maybe even coypu, armadillos and porcupines, too.

Much the same can be said for Staffordshire's resident population of big cats: clearly, these stealthy, four-legged entities have no actual business roaming around either Staffordshire or the nation as a whole. Yet, roaming around they most assuredly are – and quite contentedly, too, the evidence would seem to strongly suggest. With respect to these particular animals, the mystery is not what they are. Rather, as we have seen time and time again, the mystery is how they came to claim a stake in the Staffordshire landscape in the first place.

But, on the other hand, as we have also seen, there are many truly monstrous forms whose actions and activities are highlighted within the pages of *Mystery Animals of the British Isles: Staffordshire* that cannot be explained away quite so easily – if at all, even. Take Bigfoot, for example. The idea that a whole colony of flesh-and-blood Sasquatch-style entities might be roaming around the wooded areas of the Cannock Chase is, frankly, manifestly absurd. And the idea that they could have existed for so long without being captured, seen stealing food, slaughtering the occasional deer and more, is equally absurd in the extreme. In other words, the Staffordshire Bigfoot is far more likely to be a beast of definitively paranormal or supernatural origins (however we define what those terms may or may not mean or imply). And, very much the same can likely be said for Staffordshire's werewolves, gargoyles, pterosaurs, phantom black dogs, hairy trolls and the rest of their unearthly kind.

In other words, therefore, it seems that we have two things wildly afoot within darkest

Staffordshire: one of them involves animals of a *real* nature and one is focused around creatures that can only be described as definitively *unreal*; but both having profound effects upon those that encounter and interact with them. And, when carefully combined and mixed into one heady brew, they make for a fine, fantastic and memorable collection of creatures that only add to the wonder and the mystery that can be found throughout the county of Staffordshire.

And, speaking personally as a proud and former resident of the area, I sincerely hope that this is the way it will always be.

The Spotter's Guide

I have long thought that books should not be just a static display to be read just the once and left to pass the rest of their lives sad on a book case gathering dust to the end of their days. To me a book is a living depository of knowledge that is there to be used and abused and lives a short happy life out in the countryside. I remember as a little lad taking my father's old aircraft recognition book on trips out to the local air shows in the 1970s and '80s, and boy does that copy show it now, with its spine shattered and pages falling out*.

You might wonder the point of that. Well it is simply this, I had great fun looking up into the skies watching those old aircraft fly by (and occasionally crashing it has to be said). All the time I would be eagerly flicking through that tattered old book looking for the last plane that had flown by or seeing what plane I would like to see go by. It brought the world of vintage aircraft to life, and with that in mind I would like to present you with this, the spotter's guide to the *Mystery Animals of Staffordshire*.

After all, if you are out and about in Staffordshire why not take this book with you, you never know what you might stumble across.

* You will be glad to know while the original book died happy another copy was bought some years later to replace it.

Glen Vaudrey

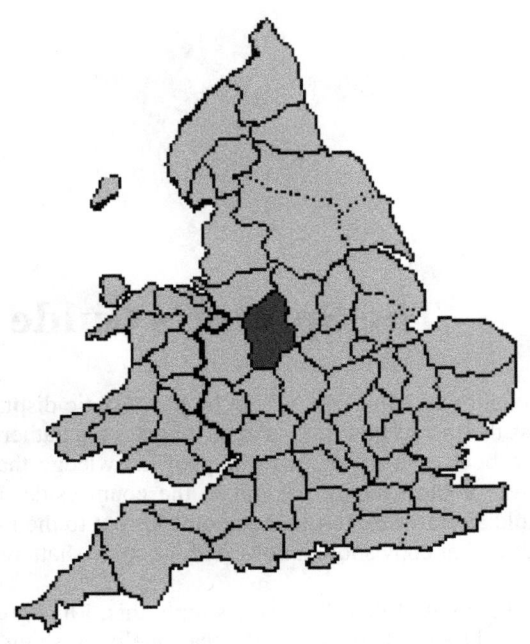

Contents of the Spotter's Guide

Big cat
Bigfoot
Black dog
Crocodile
Gabriel hounds
Jenny Greenteeth
Mermaid
Phantom animal dumper
Taigheirm
Vampire
Werewolf
White rabbit
Winged creature

Big Cat

What is it?
A large mystery feline, one of many sightings reported in the United Kingdom each year.

Where has it been seen?
Cannock Chase, Stafford and Tamworth are just a few of the locations where a big cat has been seen in recent years. But sightings are just as likely elsewhere in the county so keep an eye open for one.

What to look for
Try a big cat far in excess of the average domestic cat, for example if the cat in question was a melanistic leopard you could be looking at one measuring up to 6½ feet in length and weighing up to 200lbs, that's far bigger than any tubby tabby cat that might later be produced to explain away a sighting. You are more likely to come across paw marks or scratch marks on trees than an actual cat, while not as impressive as a sighting of the animal they are still important evidence in the quest for identifying this unknown cat.

What to do if you see one
If it is safe take a picture of it. It is advisable not to approach the animal but rather just observe from where you are, still it might be worth keeping your fingers crossed that it heads away from you; most reports suggest that they will leave with no interaction. Once the animal has gone it might be worthwhile to take a few photographs of any prints.

Worth considering
It is surprisingly easy to confuse a large domestic cat in the quest for this mystery cat, especially if the sighting is at a distance it can become hard to gauge the animal's size.

Had a sighting?
Lucky you; fill in your notes here and please let the folks at CFZ know.

Bigfoot

What is it?
A large ape-like man.

Where has it been seen?
In 1879 in the area around bridge 39 on the Shropshire Canal, since then a number of ape-men have made appearances in the county causing a great deal of unease to those who have had a close encounter.

What to look for
An unusual monkey-like ape-man that appears out of nowhere.

What to do if you see one
Stagger back in amazement.

Worth considering
It's not just a bloke in a gorilla suit on his way to a fancy dress party.

Had a sighting?
When you've calmed down, put your notes here.

Black Dog

<u>What is</u> <u>it?</u>
A good question, it looks like a dog, a very big dog in fact.

<u>Where has it been seen?</u>
There are plenty of places in the county where you could encounter a mysterious dog, in Brereton, on Cannock Chase, or more likely that most haunted of locations Kidsgrove.

<u>What to look for</u>
A strange mysterious hound that seems to have appeared out of nowhere and disappears equally mysteriously.

<u>What to do if you see one</u>
Keep your fingers crossed it isn't looking for you, after all they are either portents of impending doom, or at the very least scary looking creatures.

<u>Worth considering</u>
It could be a real dog and not some phantom hound of doom, so no need to worry.

<u>Had a sighting?</u>
Well if you are a miner going to the pit it's probably a good time to turn around and head home. If you're not a miner write your notes here.

Crocodile

What is it?
A dirty great big reptile that you wouldn't want to go swimming with.

Where has it been seen?
Back in 2003 it appeared that the waters around Cannock were playing host to this creature.

What to look for
A strange floating log that seems to be watching you, and a lack of ducks.

What to do if you see one
For a start don't go swimming in the water with it.

Worth considering
It is a log, and the ducks are just elsewhere.

Had a sighting?
Make your notes here.

Gabriel Hounds

What is it?
A musical heavenly host racing across the heavens.

Where has it been seen?
Not seen but rather heard, up in the heavens high above Staffordshire.

What to look for
Not as much look for but listen out for a sound more melodious than you have ever heard descending from the sky.

What to do if you see one
Sit back and relax and listen to a heavenly music.

Worth considering
It might not be a bunch of happy singing hounds racing across the sky, alternatively it could be a wild hunt chasing down a lost soul.

Had a sighting?
I hope you were able to record some of the music you heard.

Jenny Greenteeth

What is it?
A strange underwater woman with green skin, hair and teeth. Fair to say she isn't a looker.

Where has it been seen?
She is rumoured to inhabit any duckweed-covered stretch of water.

What to look for
A vast mat of duckweed calmly floating on the surface of the water in a lonely looking place, maybe just the odd eye to be spotted amongst the greenery.

What to do if you see one
Don't go too near the water's edge for a start or Jenny will be the last thing you will ever see.

Worth considering
It could just be a mat of duckweed, still it's worth keeping away from the edge just in case.

Had a sighting?
If so make your notes here.

Mermaid

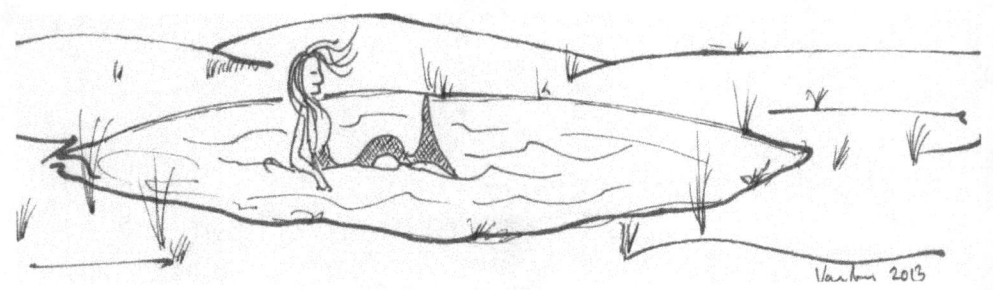

What is it?
Staffordshire's very own inland half-fish half-woman creature.

Where has it been seen?
The Blakemere pool near the village of Thorncliffe.

What to look for
A beautiful looking mermaid sitting combing her long hair at the water's edge in the dead of night.

What to do if you see one
Best keep clear, for despite all that hair combing she's a dangerous creature only trying to pull you under the water.

Worth considering
Not to be confused with the nearby pub which, while it is called The Mermaid, isn't actually one.

Had a sighting?
Make your notes here.

Phantom Animal Dumper

What is it?
Many strange out-of-place animals have turned up in Staffordshire over the years, ranging from armadillos, wild boar, pythons, coypu and wallabies. No one knows where they have come from but the finger of blame is often pointed in the direction of a man in a van dumping them in the countryside.

Where has it been seen?
The point is it has never been seen, only heard about.

What to look for
A big van parked in a lay-by with lots of animal sounds coming from it.

What to do if you see one
Get your camera out and get taking some photos.

Worth considering
It might be that equally fictitious driver who dumps urban foxes in the countryside.

Had a sighting?
You must be kidding, no one has ever seen this fellow in action. But if you have seen him at work write down his number plate here.

Taigheirm

What is it?
A shocking cat bonfire, intended to call up the most amusingly named daemon cat.

Where has it been seen?
Deepest darkest Staffordshire, and long ago on the Isle of Mull.

What to look for
A group of hooded figures acting rather shifty near a bonfire.

What to do if you see one
Stand back in shock and horror that such things are going on. Saying that, it might be worth hanging around to see if Big Ears the daemon cat turns up. On second thoughts running away might be a better idea.

Worth considering
It could be just a bunch of clean living folk keeping warm by a big bonfire.

Had a sighting?
Rather disturbing to witness wasn't it.

Vampire

What is it?
Traditionally a blood sucking undead creature.

Where has it been seen?
Well at least one person thought it could be found in Stoke in the 1970s.

What to look for
Hard to say really as the chief witness died without giving too much away about the creature he thought was hunting him.

What to do if you see one
Keep away from it, not because it's a blood-sucking creature of the night, but rather because they seem to have a liking for eating human excrement which isn't all that pleasant when you think about it.

Worth considering
It's just a weird pasty looking bloke you've spotted.

Had a sighting?
Get those bags of salt out and crack open the garlic.

Werewolf

<u>What is it?</u>
A real life wolf man just like you see in the horror films.

<u>Where has it been seen?</u>
Once again you need to head to Cannock Chase.

<u>What to look for</u>
A tall hairy wolf-like creature walking on two legs.

<u>What to do if you see one</u>
Jump back in amazement and fight the temptation to offer it some dog food. Instead try and take a picture of it.

<u>Worth considering</u>
It might be a big dog showing off its trick of walking on its back legs.

<u>Had a sighting?</u>
Make your notes here.

White Rabbit

What is it?
A ghostly vessel of vengeance, or alternatively just a small fluffy revenant marking the place of a murder.

Where has it been seen?
That centre of weird phantom creature sightings, Kidsgrove, as well as historically in Etruria.

What to look for
A white bunny that seems to appear and vanish seemingly at will.

What to do if you see one
Gasp in amazement having seen one of strangest beasts of doom.

Worth considering
It's just a pet white rabbit that's escaped.

Had a sighting?
Hope that your conscience is clear as it heads off hunting the dammed, if it is take a photo and make some notes.

Winged Creature

What is it?
A strange winged creature that isn't a bird and looks a little too much like a flying dinosaur for comfort.

Where has it been seen?
Bagot's Wood near Abbots Bromley.

What to look for
A strange black coloured beast that for all the world looks like a pterodactyl.

What to do if you see one
Popular opinion suggests running away. However, why not observe the creature, make some notes and take a picture.

Worth considering
It's nothing more than a common bird you haven't seen before, albeit a large and odd looking one.

Had a sighting?
If you aren't one of the folk who run away from it make your notes here.

Nick Redfern

About the Author

Nick Redfern is the author of many books on Cryptozoology, UFOs, and Forteana, including *Wildman!*; *Man-Monkey*; *Space Girl Dead on Spaghetti Junction*; *Monsters of Texas* (with Ken Gerhard); *Monster Files*; *Monster Diary*; *Memoirs of a Monster Hunter*; *There's Something in the Woods*; *Contactees*; *Final Events*; *The Real Men in Black*; *The NASA Conspiracies*; *Science Fiction Secrets*; *On the Trail of the Saucer Spies*; and *Strange Secrets*.

Nick has written for Britain's *Daily Express* newspaper, *Military Illustrated*, and *Penthouse* magazine.

He has appeared on more than 70 TV shows, including: Fox News; the BBC's Out of This World; the SyFy Channel's Proof Positive; the Space Channel's Fields of Fear; the History Channel's Monster Quest, America's Book of Secrets, Ancient Aliens and UFO Hunters; the National Geographic Channel's Paranatural; and MSNBC's Countdown with Keith Olbermann.

Originally from the UK, Nick lives on the fringes of Dallas, Texas. He likes his Carlsberg Special Brew and Punk Rock and can be contacted at: http://nickredfernfortean.blogspot.com

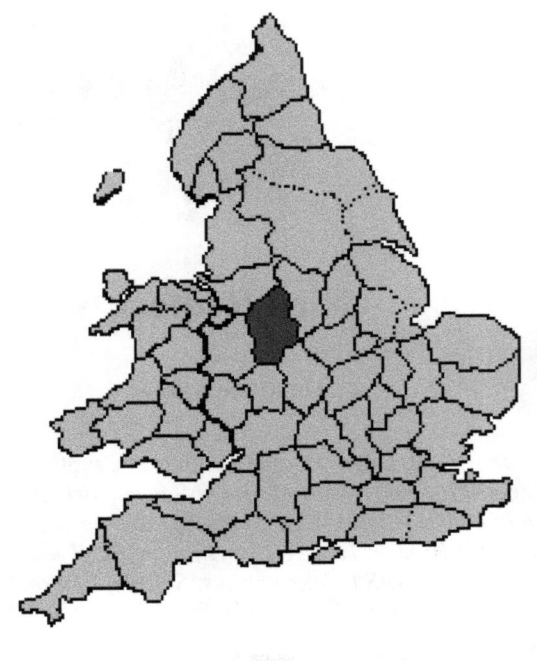

References and Resources

Chapter I: *Beware the Black Dog*
Explore Phantom Black Dogs, Bob Trubshaw, Heart of Albion Press, 2005
www.cannock-chase.net
The Black Dog of Bungay – A Brief History, http://www.bungay-suffolk.co.uk
Interviews with Nigel Lea, May 5 and May 7, 2006
Scary Encounters with Phantom Black Dogs, http://www.squidoo.com/who-let-the-dogs-out
Did They See the 'Ghost Dog' of Brereton? Cannock Advertiser, January 18, 1985
Interview with Sally Armstrong, November 22, 2000
One Ghost and His Dog, Lichfield Post, February 2, 1995
Wolf-Like Creature Terrifies Motorists, Chase Post, June 28, 2006
Great Beast Debate on Net, Chase Post, July 6, 2006
Interview with Jim Broadhurst, October 5, 2009
Interview with the Bradley family, September 21, 2009
Interview with Marjorie Saunders, August 22, 2009
Interview with Gerald Clarke, December 23, 2000
Strange Staffs, Tim Prevett, *Paranormal*, February 2009
http://lichfield-cathedral.org
http://en.wikipedia.org/wiki/Swythamley_Hall
Fresh Sighting of UFOs and Werewolves on Cannock Chase, Birmingham Post, January 15, 2010

Chapter II: *Water Maidens*
The Legendary Lore of the Holy Wells of England, Including Rivers, Lakes, Fountains and Springs, Robert Charles Hope, Elliot Stock Books, 1883
The History, Gazetteer and Directory of Staffordshire, 1851
http://www.antipope.org
The Blakemere Mermaid of Morridge and Mermaids of Other Places, Lisa Dowley, *The Centre for Fortean Zoology Yearbook, 2007*, CFZ Press, 2007
History of the Ancient Parish of Leek, John Sleigh, 1883
Interview with John Davis, June 22, 1989
Interview with family, October 30, 2000
Time-Slips: Doorways to the Past or Future, Nick Redfern, *Paranormal*, Issue 38, August 2009
Battle of Hopton Heath, http://en.wikipedia.org/wiki/Battle_of_Hopton_Heath

Chapter III: *Out-of-Place Cats*
Parliamentary copyright material from Commons *Hansard*, February 2, 1998, is reproduced with the permission of the Controller of Her Majesty's Stationery Office on behalf of Parliament.
Documentation secured under the terms of the British Government's Freedom of Information Act.

Chapter IV: *Creatures on the Loose*
Interview with source, June 14, 2006
Interview with source, October 3, 2000
Here, Puss! Puss! Puss!, *Chase Post*, January 2000
Chase Beast's Getting Bolder, *Chase Post*, March 2, 2000
Man-Monkeys and Big Cats, *Chase Post*, August 31, 2000
Cannock Chase – German Military Cemetery, www.cannockchasedc.gov.uk
Paws for Thought over Cat Sightings, *Chase Post*, March 23, 2006
Man Finds Beast's Skull, *Birmingham Post*, March 29, 2006
Mutilated Deer Discovered, *Birmingham Post*, March 29, 2006
Interview with Jack Burks, November 22, 2000

Chapter V: *Staffordshire Goes Big-Cat Crazy!*
Chase Post, April 3, 2008
Has beast claimed a kill on the Chase?, *Chase Post*, May 8, 2008
Chase Post, May 22, 2008
Sighting adds to Panther Mystery, *Stafford Post*, June 5, 2008
My Encounter with the Beast, *Stafford Post*, October 30, 2008
Expert Plays Down Threat as Chase Panther Spotted Close to Homes, *Chase Post*, July 23, 2009
Tamworth Herald, July 30 2009
Tamworth Herald, September 28, 2009
Tamworth Herald, October 19, 2009
Tamworth Herald, November 5, 2009

Chapter VI: *From Where do They Come?*
I Helped Raise Chase Big Cat, *Birmingham Post*, March 29, 2006
Where did all the Doctor's Pets Go?, *Chase Post*, May 3, 2006
Big Cat Territory in our Midst, www.shropshirestar.com
http://www.chillingtonhall.co.uk
http://en.wikipedia.org/wiki/Chillington_Hall
Big Cats on the Prowl, *BBC Inside Out*, February 10, 2003

Chapter VII: The Taigheirm Terror
The Owlman and Others: 30th Anniversary Edition, Jonathan Downes, CFZ Press, 2006
The Mystery Animals of the British Isles: The Western Isles, Glen Vaudrey, CFZ Press, 2009
Interview with Eileen Allen, June 4, 2009
Interviews with Bob Parker, January 8 and 12, 2001

Interview with Sally Ward, May 29, 2009
Interviews with Donald Johnson, May 23, 2009, August 12 and 31, 2009
Mystery Big Cats, Merrily Harpur, Heart of Albion, 2006
Supernatural Sacrifice: Big Cats and the Paranormal, Donald Johnson, publication-pending
The Tiger in the House, Carl Van Vechten, A.A. Knopf, 1922
The History of the Prince, the Lords Marcher, and the Ancient Nobility of Powys Fadog and the Ancient Lords of Arwystli, Cedewen, and Meirionydd, J.Y. W. Lloyd, T. Richards, 1881

Chapter VIII: *Creatures of the Castle Ring*
www.cannock-chase.net
Interviews with Pauline Charlesworth, June 19, 2001 and August 13, 2009
Cannock Mercury, October 1995
Interviews with Alec Williams, June 19, 2004 and August 15, 2009
Hunt for Dark Forces at Chase Monument, *Chase Post*, June 8, 2005
Paranormal Team Report UFO Activity, *Chase Post*, September 21, 2005
Group's Spooky Findings, *Birmingham Post*, February 15, 2006

Chapter IX: *Where the Bigfoot Lurks*
http://en.wikipedia.org/wiki/Bigfoot
http://en.wikipedia.org/wiki/Bigfoot#Gigantopithecus
http://en.wikipedia.org/wiki/Almasty
http://en.wikipedia.org/wiki/Yowie
http://en.wikipedia.org/wiki/Yeti

Man-Monkey: In Search of the British Bigfoot, Nick Redfern, CFZ Press, 2007
Wildman! A Guide to British Man-Beasts, Nick Redfern, to be published by the CFZ in 2010
Shropshire Folklore, Charlotte S. Burne & Georgina F. Jackson, Trubner, 1883
Interviews with Barry and Elaine, May 28, November 1 and December 3, 2000 and April 19, 2009
Interviews with Mick Dodds, February 17, 2001 and January 27, 2009
www.castleuk.net
Interviews with Jackie Houghton, May 4, 2001 and November 28, 2009
Interview with Gavin Addis, May 5, 2001
www.walkingbritain.co.uk
www.roman-britain.org
www.staffordshire.gov.uk
www.gcbro.com
Night Terror with a British Bigfoot, Peter Rhodes, *Express & Star*, January 11, 2003
The Morning Show, BBC, 2003
http://news.bbc.co.uk/1/shared/spl/hi/programmes/morning_show/html/myths.htm
Interview with Tom, June 19, 2006

Chapter X: *Sasquatch Mania*
Bigfoot in Britain, *Fate*, January 2006
Birmingham Mail, February 14, 2006

Chase Post, February 9, 2006
Bigfoot or Big Cat – Frenzy Continues, *Chase Post*, February 22, 2006
Chase Post, February 29, 2006
Bigfoot Almost Made me Lose my Baby, *Chase Post*, March 23, 2006
Email, May 17, 2009
The Little-Known History of the Hill, *Lichfield Post*, January 15, 2010

Chapter XI: *There's something in the Water*
Field Report: The Cannock Crocodile, Jonathan Downes, www.cfz.org.uk, undated
Mystery as 'croc' spotted at pool, *Wolverhampton Express and Star*, June 16, 2003
There is something Very Fishy in Pool!, *Chase Post*, October 1, 2009
Interview with Norman Dodd, August 29, 1995
Surprise Catch in Canal Turns out to be 14ft Burmese Python, *Birmingham Evening Mail*, January 24, 2003
Chase Post, June 2006
Major Probe into Cannock Chase Mysteries, *Chase Post*, June 15, 2006
Author Returns to Investigate Area's Unexplained Mysteries, *Walsall Chronicle*, June 22, 2006
Snakes, Tarantulas, Crocodiles, Panthers – on the Chase!, *Birmingham Post*, September 20, 2006

Chapter XII: *Monsters of the Full Moon*
The Wolf Man, Universal Studios, 1941
Lycanthropy: New Evidence of its Origin, H.F.Moselhy, *Psychopathology*, Vol. 32, issue 4, July-August, 1999
http://en.wikipedia.org/wiki/Peter_Stumpp
www.paranormal.lovetoknow.com
www.shanmonster.com/witch/werewolf/chalons.html
Hunting the American Werewolf, Linda Godfrey, Trails Media Group, 2006
Interview with Margaret Shelley, October 3, 2008
Interview with Pat Shirley, March 4, 2003 and September 3, 2009
Interview with Sid Lavender, December 3 and December 9, 2009
Interview with the Jacoby family, December 12, 2009
BBC Nationwide, 1976
The Evening Chronicle, January 23, 2003
There's something in the Woods, Nick Redfern, Anomalist Books, 2008
http://en.wikipedia.org/wiki/Alrewas
Stafford Post, April 26, 2007
"Underground Tribe" Theory on Chase Beast, *Chase Post*, May 14, 2007
Chase Search for Weird and Wonderful, *Stafford Post*, June 13, 2007
Wolves May Live on Chase – Expert, *Chase Post*, May 30, 2007
http://enwikipedia.org/wiki/SS; http://enwikipedia.org/Maximillian_von_Herff
Email, May 17, 2007
Email, April 5, 2007
Email, April 9, 2007

Chapter XIII: *Wallaby Wonders*
http://en.wikipedia.org/wiki/Wallaby
Monster Hunter, Jonathan Downes, CFZ Press, 2004
Interview with Alice Morris, October 18, 2009
Strewth! It's an Albino Wallaby, Daily Mail, May 30, 2007
http://en.wikipedia.org/wiki/Peak_District
Return of the Wallabies, Daily Mail, July 9, 2009
Modern Life kills off the Last of Britain's Wild Wallabies, Independent, October 9, 2000
http://en.wikipedia.org/wiki/Swythamley_Hall
http://en.wikipedia.org/wiki/The_Roaches

Chapter XIV: *Boar to be Wild*
http://en.wikipedia.org/wiki/Wild_boar
http://en.wikipedia.org/wiki/Wild_boar#Status_in_Britain
http://en.wikipedia.org/wiki/Wild_boar#Mythology.2C_religion.2C_history_and_fiction
http://www.defra.gov.uk/news/2008/080219b.htm
Interview with Colin Blakemore, February 14, 2009
Interview with Gareth and Dan, October 29, 2007
http://en.wikipedia.org/wiki/Coypu
www.wildaboutbritain.co.uk/coypu
www.nonnativespecies.org
Hansard, December 21, 1965
Time, September 1, 1961
Interview, August 3, 2000

Chapter XV, *Alien Animals*:
http://en.wikipedia.org/wiki/Porcupine
http://news.bbc.co.uk/2/hi/uk_news/england/gloucestershire/4166627.stm
http://en.wikipedia.org/wiki/Armadillo
Interview with Muriel Atkins, September 8, 2000
Snow May Solve the Pine Marten Puzzle, Stafford Post, January 13, 2010
Cannock Chase is hit by Muntjac Mayhem, Chase Post, January 15, 2010

Chapter XVI, *Flying Monsters*:
Interviews with Pauline Charlesworth, June 6, 2001 and May 6, 2008
Monster!, Neil Arnold, CFZ Press, 2007
The Mothman Prophecies, John Keel, Saturday Review Press, 1975
The Owlman and Others: 30th Anniversary Edition, Jonathan Downes, CFZ Press, 2006
Memoirs of a Monster Hunter, Nick Redfern, New Page Books, 2007
Monster! Neil Arnold, CFZ Press, 2007
The Brentford Griffin, Andy Collins, Earthquest, 1985
Humanoid Contact: The Cases 2004, www.thelosthaven.co.uk
Chase Post, February 19, 2009
Interviews with Alfred Tipton January 9, 2007 and November 15, 2009
http://www.nationalforest.org/visit/index.php?fuseaction=location.showlocation&loc_id=125

http://en.wikipedia.org/wiki/Needwood_Forest
http://www.blithfieldhall.co.uk
http://en.wikipedia.org/wiki/Blithfield_Hall
Giant Moth Wings way to Tamworth, Tamworth Herald, January 14, 2010

Acknowledgements

I would like to offer my very sincere thanks and appreciation indeed to all of those people who were willing to share with me the details of their own, personal encounters with the many and varied unknown and out-of-place animals of Staffordshire – without you, this book could never have been written.

I would also like to offer a very special thank-you to the following people, equally without whom this book would not have seen the light of day: my good friend Jonathan Downes, for encouraging me to write *The Mystery Animals of the British Isles: Staffordshire*; Jon's distinctly better-half, Corinna, and everyone else at the Centre for Fortean Zoology, and particularly Richard Freeman, Graham Inglis and Ollie Lewis; Mark North, for a variety of stories sent my way; all of the staff at the *Chase Post* and especially so the newspaper's editor Mike Lockley, for flying the flag of Cannock's high-strangeness and for giving me permission to cite the newspaper's many and varied articles on weird and wonderful creatures; and good friend and author Neil Arnold.

Nick Redfern

Glen Vaudrey

About the Author

Glen Vaudrey was born into a dying farming community in Lancashire in 1972, the eldest son of farm labourers. He spent a decade travelling the more remote parts of northern and Eastern Europe during which time he worked variously as a court usher, regularly cooked breakfast for 70, farmed snails and at one low point was to be found cleaning public toilets. Not just an author of mystery animal books but an artist with a strong worldwide following. He currently lives in the Cheshire village deserted by his ancestors over a hundred years ago, hopefully whatever drove them out of the village has long been forgotten. And he revels in the title of natural mystic.

Bibliography

Hippisley Coxe, Antony D. *Haunted Britain* (Hutchinson of London, 1973)
Leese, Philp R. *The Kidsgrove Boggart and the Black Dog* (Staffordshire Library, Arts & Archives, 1989)
Pye, John *The Force Was With Me* published in 2005
Readers Digest *Folklore, Myths and Legends of Britain* (Readers Digest, 1973)
Travis, Peter *In Search of the Supernatural* (Wolfe Publishing Limited, 1975)
Trubshaw, Bob *Explore Phantom Black Dogs* (Heart of Albion, 2005)
Westwood Jennifer, Simpson Jacqueline *The Lore of the Land* (Penguin Books, 2005)
Woodward, Ian *The Werewolf Delusion* (Paddington Press, 1979)

The following website has been consulted and is to be highly recommended
www.paranormaldatabase.com

Acknowledgements

I would like to thank John Pye for his information regarding the Villas Vampire; thanks also to The Mysteries Group for the warm welcome they gave me when I visited them in Stoke.

Glen Vaudrey

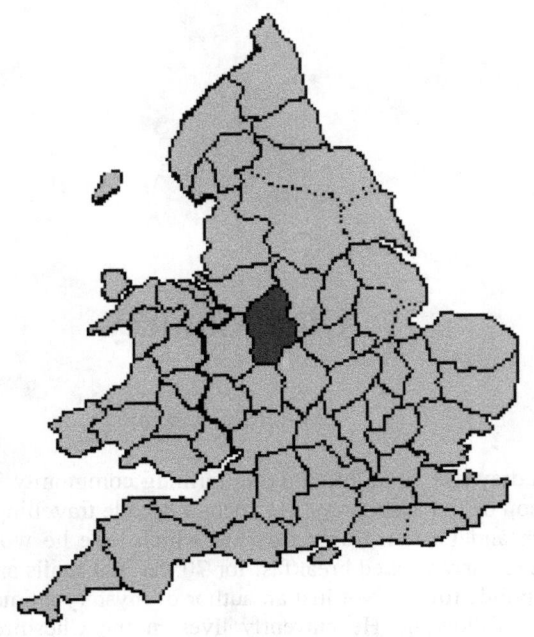

INDEX

Abominable Snowman 81
Alien Big Cats (ABCs) *see* big cats
alligators 114
   see also crocodiles
Almas 81
Alrewas 132–133, 152
Alton Towers 22
Aqualate Mere 35
armadillos 111, 153, 171

Baddesley Common 57
Bagot's Wood 154, 161, 162–163
Beaudesert Old Park 77
big cats 37–40, 41–50, 51–57, 99–101, 171, 176
   Cannock 41
   Cannock Chase 42–50, 51–54, 61, 66, 92, 93
   Mull 70, 71
   origins 56–63
   RAF Stafford 46–47
   Stafford 53
   supernatural 65–66, 72, 101
   Tamworth 55–57
   see also Taigheirm
Bigfoot 81–82, 89, 130, 171, 177
   Cannock Chase 85–88, 91–95
   Man-Monkey 83
   media coverage 91–95
birds 124
   see also flying creatures
black cats *see* big cats
black dogs 13–14, 97, 178
   Alton Towers 22

   Brereton 15–16
   Burntwood 16–17
   Cannock Chase 14–15, 20
   Hammerwich 20–21
   Ipstones 21
   Kidsgrove 22–27
   Leek 21
   RAF Stafford 20
   Swythamley Hall 19
   Tamworth Castle 19
   Wednesbury 21
Black Witch moths 164–165
Blakemere 32–33, 35
Blithfield Hall 161–162
boar *see* wild boar
boggart 23–24, 26–27
Brentford 160
Brereton 15–16
Brindley Heath 46
Brockton 149
Brocton 93–94
Brownhills 163
Buckinghamshire 140
Burmese Pythons 123
Burntwood 16–17

caiman 121
   see also crocodiles
Cannock 41, 113
Cannock Chase 14
   big cats 42–50, 51–54, 61, 66, 92, 93
   Bigfoot 85–88, 91–95
   black dogs 14–15, 17–18, 20
   fish 121–122

flying creatures 163–164
goblin-like creatures 84
Muntjac deer 154–155
wallabies 141–142
werewolves 128–131, 133
wild boar 148–149
wild men 104
   see also Castle Ring; German Cemetery
Cannock Wood 75, 78, 86, 149
Castle Ring 75–79, 102–103, 149
cat sacrifices *see* Taigheirm
cats *see* big cats
cavemen 129, 130
Chartley Castle 85
Chillington Hall 59
chimpanzee-like creature 85
Civil War battle 34–35
common snapping turtles 105
Cornwall 67, 159
coypu 111, 151–152, 171
Crank 168
crocodiles 113–121, 179

deer 154–155
Devon 147
Dordon 56
Dorset 147

East Sussex 147
Edgbaston 123
eels 122–123
Essex 151
Etching Hill 95–96
Etruria 168–170

fish 120, 121–122
flying creatures 76, 112, 157–160, 162–165, 188
flying man 163–164
Forest of Dean 146, 147, 152

Gabriel hounds 20–21, 180
Gentleshaw 78
German Cemetery 43–44, 65, 89–90, 107, 128–129, 131

Ghost Dog of Brereton 15–16
ghostly dogs *see* black dogs
Gigantopithecus 82
Glacial Boulder 85–86
Gloucestershire 147
goblin-like creatures 84
Great Wyrley 163
Griffin 160
Gun Hill 19

Hammerwich 20–21
Hampshire 39
Harecastle Tunnel 23
headless woman 23, 24
Hednesford 153
Herefordshire 147, 154
Hermitage Farm 21
Hopton Heath 33, 34–35

Ingestre Park Golf Club 45
Ipstones 21
Ireland 140
Isle of Man 139

Jenny Greenteeth 35–36, 181

Kent 147
Kidsgrove 22–27, 167

Lancashire 168
Leek 19, 21, 30, 133, 148
Lewis 134
Lichfield 18, 153
Loch Morar 127
Lower Ash 24
lynx 39–40, 42, 99

man-beasts *see* Bigfoot
Man-Monkey 83
mermaids 29–34, 35, 98, 182
Milford 66, 88
monkeys 60, 61, 95
Moorlands 30–33
Mothman of Point Pleasant 158
moths 164–165

Mow Cop 26
Mull 69–71
Muntjac deer 154–155

Needwood Forest 161
Newport 35
Norfolk 15, 39–40, 139, 151–152
Northicote Farm 162–163

Oban 127
Owlman 67, 159

panthers 43–44, 46, 48, 53, 61, 65, 66
paranormal creatures 76–79, 130–131, 163, 171
   big cats 65–66, 72, 101
   see also black dogs
Pars Warren 46
Peak District 140–141
Penkridge 41, 87
phantom dogs see black dogs
pine martens 153–154
Polesworth 56
porcupines 152, 171
Pterodactobats 159–160
pterosaurs 162–163
pumas 56, 68
Pye Green 48
pythons 123

rabbits 167–170, 187
RAF Alconbury 131–132
RAF Stafford 20, 46–47
Ranton 83
redtail hawks 124
Roaches, The 140–141
Roman View Pool 105, 113–121
Ross on Wye 147
Rugeley 15–16, 95

s sightings 78
Sasquatch see Bigfoot
Scotland
   Taigheirm 68–71
   wallabies 139

werewolves 127, 134
wild boar 146
serpents 122–123
Sherwood Forest 130
Shropshire 30, 154
Shugborough Hall 88–89
Slittingmill 83–84, 85, 122
snakes 122–123
snapping turtles 105
spectacled caiman 121
spectral dogs see black dogs
Spring Slade Lodge 45, 52
Stafford 53, 87
Stoke-on-Trent 135–137, 168
Suffolk 13–14, 151
supernatural creatures 76–79, 130–131, 163, 171
   big cats 65–66, 72, 101
   see also black dogs
Sussex 139
Swythamley Hall 19

Taigheirm 68–73, 184
Tamworth 55–57, 164
Tamworth Castle 19
Thorncliffe 30
Trent and Mersey Canal 22–23, 24
troll-like creatures 84
turtles 105

UFOs 158

vampires 134–137, 185

wallabies 31, 109, 139–143, 171
Walton 55
Warton 56–57
Warwickshire 154
water-maidens see mermaids
Wednesbury 21
werewolves 108, 125–131, 186
white rabbits 167–170, 187
wild boar 110, 145–149
wild men 104
Wimblebury 121–122

winged beasts 76, 112, 157–160, 162–165, 188
Wolverhampton 162–163
wolves 17–18, 106, 130, 131, 133
Woodhouses 17
Worcestershire 154

Yeti *see* Bigfoot
Yowie 81

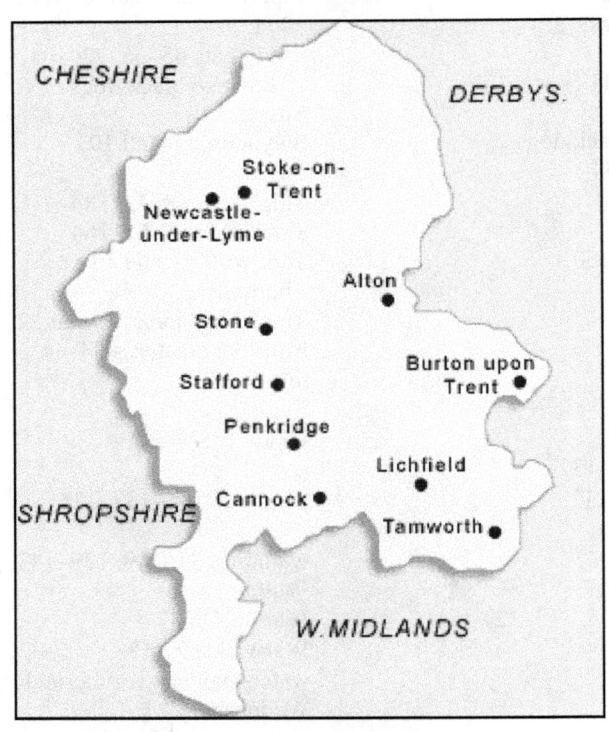

STILL ON THE TRACK OF UNKNOWN ANIMALS

T he Centre for Fortean Zoology, or CFZ, is a non profit-making organisation founded in 1992 with the aim of being a clearing house for information, and coordinating research into mystery animals around the world.

We also study out of place animals, rare and aberrant animal behaviour, and Zooform Phenomena; little-understood "things" that appear to be animals, but which are in fact nothing of the sort, and not even alive (at least in the way we understand the term).

Not only are we the biggest organisation of our type in the world, but - or so we like to think - we are the best. We are certainly the only truly global cryptozoological research organisation, and we carry out our investigations using a strictly scientific set of guidelines. We are expanding all the time and looking to recruit new members to help us in our research into mysterious animals and strange creatures across the globe.

Why should you join us? Because, if you are genuinely interested in trying to solve the last great mysteries of Mother Nature, there is nobody better than us with whom to do it.

Members get a four-issue subscription to our journal *Animals & Men*. Each issue contains nearly 100 pages packed with news, articles, letters, research papers, field reports, and even a gossip column! The magazine is Royal Octavo in format with a full colour cover. You also have access to one of the world's largest collections of resource material dealing with cryptozoology and allied disciplines, and people from the CFZ membership regularly take part in fieldwork and expeditions around the world.

The CFZ is managed by a three-man board of trustees, with a non-profit making trust registered with HM Government Stamp Office. The board of trustees is supported by a Permanent Directorate of full and part-time staff, and advised by a Consultancy Board of specialists - many of whom are world-renowned experts in their particular field. We have regional representatives across the UK, the USA, and many other parts of the world, and are affiliated with

You'll find that the people at the CFZ are friendly and approachable. We have a thriving forum on the website which is the hub of an ever-growing electronic community. You will soon find your feet. Many members of the CFZ Permanent Directorate started off as ordinary members, and now work full-time chasing monsters around the world.

Write to us, e-mail us, or telephone us. The list of future projects on the website is not exhaustive. If you have a good idea for an investigation, please tell us. We may well be able to help.

We are always looking for volunteers to join us. If you see a project that interests you, do not hesitate to get in touch with us. Under certain circumstances we can help provide funding for your trip. If you look on the future projects section of the website, you can see some of the projects that we have pencilled in for the next few years.

In 2003 and 2004 we sent three-man expeditions to Sumatra looking for Orang-Pendek - a semi-legendary bipedal ape. The same three went to Mongolia in 2005. All three members started off merely subscribers to the CFZ magazine. Next time it could be you!

We have no magic sources of income. All our funds come from donations, membership fees, and sales of our publications and merchandise. We are always looking for corporate sponsorship, and other sources of revenue. If you have any ideas for fund-raising please let us know. However, unlike other cryptozoological organisations in the past, we do not live in an intellectual ivory tower. We are not afraid to get our hands dirty, and furthermore we are not one of those organisations where the membership have to raise money so that a privileged few can go on expensive foreign trips. Our research teams, both in the UK and abroad, consist of a mixture of experienced and inexperienced personnel. We are truly a community, and work on the premise that the benefits of CFZ membership are open to all.

Reports of our investigations are published on our website as soon as they are available. Preliminary reports are posted within days of the project finishing.

Each year we publish a 200 page yearbook

We have a thriving YouTube channel, CFZtv, which has well over two hundred self-made documentaries, lecture appearances, and episodes of our monthly webTV show. We have a daily online magazine, which has over a million hits each year.

Each year since 2000 we have held our annual convention - the Weird Weekend. It is three days of lectures, workshops, and excursions. But most importantly it is a chance for members of the CFZ to meet each other, and to talk with the members of the permanent directorate in a relaxed and informal setting and preferably with a pint of beer in one hand. Since 2006 - the Weird Weekend has been bigger and better and held on the third weekend in August in the idyllic rural location of Woolsery in North Devon.

Since relocating to North Devon in 2005 we have become ever more closely involved with other community organisations, and we hope that this trend will continue. We have also worked closely with Police Forces across the UK as consultants for animal mutilation cases, and we intend to forge closer links with the coastguard and other community services. We want to work closely with those who regularly travel into the Bristol Channel, so that if the recent trend of exotic animal visitors to our coastal waters continues, we can be out there as soon as possible.

Apart from having been the only Fortean Zoological organisation in the world to have consistently published material on all aspects of the subject for over a decade, we have achieved the following concrete results:

• Disproved the myth relating to the headless so-called sea-serpent carcass of Durgan beach in Cornwall 1975
• Disproved the story

of the 1988 puma skull of Lustleigh Cleave
- Carried out the only in-depth research ever into the mythos of the Cornish Owlman.
- Made the first records of a tropical species of lamprey
- Made the first records of a luminous cave gnat larva in Thailand
- Discovered a possible new species of British mammal - the beech marten
- In 1994-6 carried out the first archival fortean zoological survey of Hong Kong
- In the year 2000, CFZ theories were confirmed when a new species of lizard was added to the British List
- Identified the monster of Martin Mere in Lancashire as a giant wels catfish
- Expanded the known range of Armitage's skink in the Gambia by 80%
- Obtained photographic evidence of the remains of Europe's largest known pike
- Carried out the first ever in-depth study of the ninki-nanka
- Carried out the first attempt to breed Puerto Rican cave snails in captivity
- Were the first European explorers to visit the `lost valley` in Sumatra
- Published the first ever evidence for a new tribe of pygmies in Guyana
- Published the first evidence for a new species of caiman in Guyana

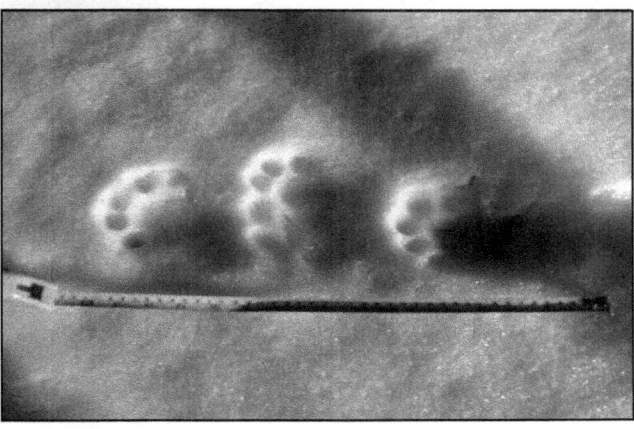

on a monster-haunted lake in Ireland for the first time
- Had a sighting of orang pendek in Sumatra in 2009
- Found leopard hair, subsequently identified by DNA analysis, from rural North Devon in 2010
- Brought back hairs which appear to be from an unknown primate in Sumatra
- Published some of the best evidence ever for the almasty in southern Russia

CFZ Expeditions and Investigations include:

- 1998 Puerto Rico, Florida, Mexico (Chupacabras)
- 1999 Nevada (Bigfoot)
- 2000 Thailand (Naga)
- 2002 Martin Mere (Giant catfish)
- 2002 Cleveland (Wallaby mutilation)
- 2003 Bolam Lake (BHM Reports)

- 2003 Sumatra (Orang Pendek)
- 2003 Texas (Bigfoot; giant snapping turtles)
- 2004 Sumatra (Orang Pendek; cigau, a sabre-toothed cat)
- 2004 Illinois (Black panthers; cicada swarm)
- 2004 Texas (Mystery blue dog)
- Loch Morar (Monster)
- 2004 Puerto Rico (Chupacabras; carnivorous cave snails)
- 2005 Belize (Affiliate expedition for hairy dwarfs)
- 2005 Loch Ness (Monster)
- 2005 Mongolia (Allghoi Khorkhoi aka Mongolian death worm)

- 2006 Gambia (Gambo - Gambian sea monster, Ninki Nanka and Armitage's skink
- 2006 Llangorse Lake (Giant pike, giant eels)
- 2006 Windermere (Giant eels)
- 2007 Coniston Water (Giant eels)
- 2007 Guyana (Giant anaconda, didi, water tiger)
- 2008 Russia (Almasty)
- 2009 Sumatra (Orang pendek)
- 2009 Republic of Ireland (Lake Monster)
- 2010 Texas (Blue Dogs)
- 2010 India (Mande Burung)
- 2011 Sumatra (Orang-pendek)

For details of current membership fees, current expeditions and investigations, and voluntary posts within the CFZ that need your help, please do not hesitate to contact us.

The Centre for Fortean Zoology,
Myrtle Cottage,
Woolfardisworthy,
Bideford, North Devon
EX39 5QR

Telephone 01237 431413
Fax+44 (0)7006-074-925
eMail info@cfz.org.uk

Websites:

www.cfz.org.uk
www.weirdweekend.org

THE WORLD'S WEIRDEST PUBLISHING COMPANY

HOW TO START A PUBLISHING EMPIRE

Unlike most mainstream publishers, we have a non-commercial remit, and our mission statement claims that "we publish books because they deserve to be published, not because we think that we can make money out of them". Our motto is the Latin Tag *Pro bona causa facimus* (we do it for good reason), a slogan taken from a children's book *The Case of the Silver Egg* by the late Desmond Skirrow.

WIKIPEDIA: "The first book published was in 1988. *Take this Brother may it Serve you Well* was a guide to Beatles bootlegs by Jonathan Downes. It sold quite well, but was hampered by very poor production values, being photocopied, and held together by a plastic clip binder. In 1988 A5 clip binders were hard to get hold of, so the publishers took A4 binders and cut them in half with a hacksaw. It now reaches surprisingly high prices second hand.

The production quality improved slightly over the years, and after 1999 all the books produced were ringbound with laminated colour covers. In 2004, however, they signed an agreement with Lightning Source, and all books are now produced perfect bound, with full colour covers."

Until 2010 all our books, the majority of which are/were on the subject of mystery animals and allied disciplines, were published by `CFZ Press`, the publishing arm of the Centre for Fortean Zoology (CFZ), and we urged our readers and followers to draw a discreet veil over the books that we published that were completely off topic to the CFZ.

However, in 2010 we decided that enough was enough and launched a second imprint, `Fortean Words` which aims to cover a wide range of non animal-related esoteric subjects. Other imprints will be launched as and when we feel like it, however the basic ethos of the company remains the same: Our job is to publish books and magazines that we feel are worth publishing, whether or not they are going to sell. Money is, after all - as my dear old Mama once told me - a rather vulgar subject, and she would be rolling in her grave if she thought that her eldest son was somehow in `trade`.

Luckily, so far our tastes have turned out not to be that rarified after all, and we have sold far more books than anyone ever thought that we would, so there is a moral in there somewhere...

Jon Downes,
Woolsery, North Devon
July 2010

CFZ PRESS

Other Books in Print

Sea Serpent Carcasses - Scotland from the Stronsa Monster to Loch Ness by Glen Vaudrey
The CFZ Yearbook 2012 edited by Jonathan and Corinna Downes
ORANG PENDEK: Sumatra's Forgotten Ape by Richard Freeman
THE MYSTERY ANIMALS OF THE BRITISH ISLES: London by Neil Arnold
CFZ EXPEDITION REPORT: India 2010 by Richard Freeman *et al*
The Cryptid Creatures of Florida by Scott Marlow
Dead of Night by Lee Walker
The Mystery Animals of the British Isles: The Northern Isles by Glen Vaudrey
THE MYSTERY ANIMALS OF THE BRTISH ISLES: Gloucestershire and Worcestershire by Paul Williams
When Bigfoot Attacks by Michael Newton
Weird Waters – The Mystery Animals of Scandinavia: Lake and Sea Monsters by Lars Thomas
The Inhumanoids by Barton Nunnelly
Monstrum! A Wizard's Tale by Tony "Doc" Shiels
CFZ Yearbook 2011 edited by Jonathan Downes
Karl Shuker's Alien Zoo by Shuker, Dr Karl P.N
Tetrapod Zoology Book One by Naish, Dr Darren
The Mystery Animals of Ireland by Gary Cunningham and Ronan Coghlan
Monsters of Texas by Gerhard, Ken
The Great Yokai Encyclopaedia by Freeman, Richard
NEW HORIZONS: Animals & Men issues 16-20 Collected Editions Vol. 4 by Downes, Jonathan
A Daintree Diary -
Tales from Travels to the Daintree Rainforest in tropical north Queensland, Australia by Portman, Carl
Strangely Strange but Oddly Normal by Roberts, Andy
Centre for Fortean Zoology Yearbook 2010 by Downes, Jonathan
Predator Deathmatch by Molloy, Nick
Star Steeds and other Dreams by Shuker, Karl
CHINA: A Yellow Peril? by Muirhead, Richard
Mystery Animals of the British Isles: The Western Isles by Vaudrey, Glen

Giant Snakes - Unravelling the coils of mystery by Newton, Michael
Mystery Animals of the British Isles: Kent by Arnold, Neil
Centre for Fortean Zoology Yearbook 2009 by Downes, Jonathan
CFZ EXPEDITION REPORT: Russia 2008 by Richard Freeman *et al*, Shuker, Karl (fwd)
Dinosaurs and other Prehistoric Animals on Stamps - A Worldwide catalogue by Shuker, Karl P. N
Dr Shuker's Casebook by Shuker, Karl P.N
The Island of Paradise - chupacabra UFO crash retrievals, and accelerated evolution on the island of Puerto Rico by Downes, Jonathan
The Mystery Animals of the British Isles: Northumberland and Tyneside by Hallowell, Michael J
Centre for Fortean Zoology Yearbook 1997 by Downes, Jonathan (Ed)
Centre for Fortean Zoology Yearbook 2002 by Downes, Jonathan (Ed)
Centre for Fortean Zoology Yearbook 2000/1 by Downes, Jonathan (Ed)
Centre for Fortean Zoology Yearbook 1998 by Downes, Jonathan (Ed)
Centre for Fortean Zoology Yearbook 2003 by Downes, Jonathan (Ed)
In the wake of Bernard Heuvelmans by Woodley, Michael A
CFZ EXPEDITION REPORT: Guyana 2007 by Richard Freeman *et al*, Shuker, Karl (fwd)
Centre for Fortean Zoology Yearbook 1999 by Downes, Jonathan (Ed)
Big Cats in Britain Yearbook 2008 by Fraser, Mark (Ed)
Centre for Fortean Zoology Yearbook 1996 by Downes, Jonathan (Ed)
THE CALL OF THE WILD - *Animals & Men issues 11-15 Collected Editions Vol. 3* by Downes, Jonathan (ed)
Ethna's Journal by Downes, C N
Centre for Fortean Zoology Yearbook 2008 by Downes, J (Ed)
DARK DORSET -Calendar Custome by Newland, Robert J
Extraordinary Animals Revisited by Shuker, Karl
MAN-MONKEY - In Search of the British Bigfoot by Redfern, Nick
Dark Dorset Tales of Mystery, Wonder and Terror by Newland, Robert J and Mark North
Big Cats Loose in Britain by Matthews, Marcus
MONSTER! - The A-Z of Zooform Phenomena by Arnold, Neil
The Centre for Fortean Zoology 2004 Yearbook by Downes, Jonathan (Ed)
The Centre for Fortean Zoology 2007 Yearbook by Downes, Jonathan (Ed)
CAT FLAPS! Northern Mystery Cats by Roberts, Andy
Big Cats in Britain Yearbook 2007 by Fraser, Mark (Ed)
BIG BIRD! - Modern sightings of Flying Monsters by Gerhard, Ken
THE NUMBER OF THE BEAST - *Animals & Men issues 6-10 Collected Editions Vol. 1* by Downes, Jonathan (Ed)
IN THE BEGINNING - Animals & Men *issues 1-5 Collected Editions Vol. 1* by Downes, Jonathan
STRENGTH THROUGH KOI - They saved Hitler's Koi and other stories by Downes, Jonathan
The Smaller Mystery Carnivores of the Westcountry by Downes, Jonathan
CFZ EXPEDITION REPORT: Gambia 2006 by Richard Freeman *et al*, Shuker, Karl (fwd)
The Owlman and Others by Jonathan Downes
The Blackdown Mystery by Downes, Jonathan

Big Cats in Britain Yearbook 2006 by Fraser, Mark (Ed)
Fragrant Harbours - Distant Rivers by Downes, John T
Only Fools and Goatsuckers by Downes, Jonathan
Monster of the Mere by Jonathan Downes
Dragons:More than a Myth by Freeman, Richard Alan
Granfer's Bible Stories by Downes, John Tweddell
Monster Hunter by Downes, Jonathan

CFZ Classics is a new venture for us. There are many seminal works that are either unavailable today, or not available with the production values which we would like to see. So, following the old adage that if you want to get something done do it yourself, this is exactly what we have done.

Desiderius Erasmus Roterodamus (b. October 18th 1466, d. July 2nd 1536) said: "When I have a little money, I buy books; and if I have any left, I buy food and clothes," and we are much the same. Only, we are in the lucky position of being able to share our books with the wider world. CFZ Classics is a conduit through which we cannot just re-issue titles which we feel still have much to offer the cryptozoological and Fortean research communities of the 21st Century, but we are adding footnotes, supplementary essays, and other material where we deem it appropriate.

Headhunters of The Amazon by Fritz W Up de Graff (1902)

Fortean Words

The Centre for Fortean Zoology has for several years led the field in Fortean publishing. CFZ Press is the only publishing company specialising in books on monsters and mystery animals. CFZ Press has published more books on this subject than any other company in history and has attracted such well known authors as Andy Roberts, Nick Redfern, Michael Newton, Dr Karl Shuker, Neil Arnold, Dr Darren Naish, Jon Downes, Ken Gerhard and Richard Freeman.

Now CFZ Press are launching a new imprint. Fortean Words is a new line of books dealing with Fortean subjects other than cryptozoology, which is - after all - the subject the CFZ are best known for. Fortean Words is being launched with a spectacular multi-volume series called *Haunted Skies* which covers British UFO sightings between 1940 and 2010. Former policeman John Hanson and his long-suffering partner Dawn Holloway have compiled a peerless library of sighting reports, many that have not been made public before.

Other books include a look at the Berwyn Mountains UFO case by renowned Fortean Andy Roberts and a series of forthcoming books by transatlantic researcher Nick Redfern. CFZ Press are dedicated to maintaining the fine quality of their works with Fortean Words. New authors tackling new subjects will always be encouraged, and we hope that our books will continue to be as ground-breaking and popular as ever.

Haunted Skies Volume One 1940-1959 by John Hanson and Dawn Holloway
Haunted Skies Volume Two 1960-1965 by John Hanson and Dawn Holloway
Haunted Skies Volume Three 1965-1967 by John Hanson and Dawn Holloway
Haunted Skies Volume Four 1968-1971 by John Hanson and Dawn Holloway
Haunted Skies Volume Five 1972-1974 by John Hanson and Dawn Holloway
Haunted Skies Volume Six 1975-1977 by John Hanson and Dawn Holloway
Grave Concerns by Kai Roberts

Police and the Paranormal by Andy Owens
Dead of Night by Lee Walker
Space Girl Dead on Spaghetti Junction - an anthology by Nick Redfern
I Fort the Lore - an anthology by Paul Screeton
UFO Down - the Berwyn Mountains UFO Crash by Andy Roberts
The Grail by Ronan Coghlan
UFO Warminster - Cradle of Contract by Kevin Goodman
Quest for the Hexham Heads by Paul Screeton

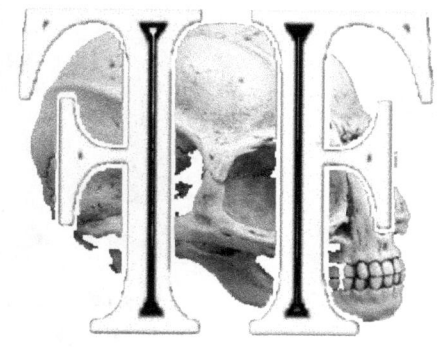

Fortean Fiction

Just before Christmas 2011, we launched our third imprint, this time dedicated to - let's see if you guessed it from the title - fictional books with a Fortean or cryptozoological theme. We have published a few fictional books in the past, but now think that because of our rising reputation as publishers of quality Forteana, that a dedicated fiction imprint was the order of the day.

We launched with four titles:

Green Unpleasant Land by Richard Freeman
Left Behind by Harriet Wadham
Dark Ness by Tabitca Cope
Snap! By Steven Bredice
Death on Dartmoor by Di Francis
Dark Wear by Tabitca Cope

www.ingramcontent.com/pod-product-compliance
Lightning Source LLC
Chambersburg PA
CBHW060512090426
42735CB00011B/2184